Modified Gravity: From Black Holes Entropy to Current Cosmology, 4th Edition

Modified Gravity: From Black Holes Entropy to Current Cosmology, 4th Edition

Guest Editor

Kazuharu Bamba

MDPI

Basel • Beijing • Wuhan • Barcelona • Belgrade • Novi Sad • Cluj • Manchester

Guest Editor
Kazuharu Bamba
Faculty of Symbiotic Systems Science
Fukushima University
Fukushima
Japan

Editorial Office
MDPI AG
Grosspeteranlage 5
4052 Basel, Switzerland

This is a reprint of the Special Issue, published open access by the journal *Entropy* (ISSN 1099-4300), freely accessible at: https://www.mdpi.com/journal/entropy/special_issues/CY8C669A5W.

For citation purposes, cite each article independently as indicated on the article page online and as indicated below:

Lastname, A.A.; Lastname, B.B. Article Title. *Journal Name* **Year**, *Volume Number*, Page Range.

ISBN 978-3-7258-7440-8 (Hbk)
ISBN 978-3-7258-7441-5 (PDF)
https://doi.org/10.3390/books978-3-7258-7441-5

Contents

About the Editor

Kazuharu Bamba

Kazuharu Bamba is a full Professor at Fukushima University in Japan. He works on inflationary cosmology, the issue of dark energy, alternative theories of gravity alongside general relativity, and the physics of black holes. He has been an Editor of various international peer-review journals such as *Entropy*, *Universe* and *Symmetry* (MDPI).

entropy

MDPI

Editorial

Modified Gravity: From Black Holes Entropy to Current Cosmology, 4th Edition

Kazuharu Bamba

Faculty of Symbiotic Systems Science, Fukushima University, Fukushima 960-1296, Japan;
bamba@sss.fukushima-u.ac.jp

Recent cosmological observational data—such as type Ia supernovae (SNe Ia) [1–3], cosmic microwave background (CMB) radiation [4–11], the cosmic large-scale structure (LSS) [12–14], baryon acoustic oscillations (BAOs) [15–18], and the effect of weak lensing [19–23]—have led researchers to confirm that, in addition to inflation in the early universe [24–27] (for recent reviews, see Refs. [28–30] and references therein), the late-time expansion of the universe is accelerating.

There are two representative ways of accounting for the late-time cosmic acceleration in the spatially flat universe. The first is to explore the so-called (unknown) dark energy with negative pressure in general relativity. The second is to introduce the extensions of general relativity at large scales, that is, to regard dark energy as a geometrical component. In regard to the second approach, a number of alternative theories of gravity have been proposed (for comprehensive reviews, see, for example, Refs. [31–71]).

This Special Issue of *Entropy*, "Modified Gravity: From Black Holes Entropy to Current Cosmology, 4th Edition", contains eight original research studies on modern astrophysics and cosmology. It is arranged as follows. In the first part, black holes [72–75] are investigated. In the second part, topics relating to cosmology [76–79] are explored. A brief overview of each article is given below.

A hot NUT–Kerr–Newman black hole is studied in Ref. [72]. The corrections for both the Hawking temperature and Bekenstein–Hawking entropy are discussed in detail through the semi-classical approach.

In Ref. [73], a hyperbolically symmetric black hole is investigated to explore how it differs from conventional classical black holes via an analysis of the dynamics of test particles around the black hole.

In Ref. [74], the modifications of entropy for charged accelerating Anti-de Sitter black holes are discussed based on correction with the gauge invariance originating from an analysis of the quantum field equations with the violations of the Lorentz symmetry for curved spacetime.

A rotating Kiselev black hole is considered in Ref. [75]. Using several approaches, such as the WKB approximation and the effect of quantum-tunneling radiation, the authors explore whether a black hole is surrounded by a quintessence field, playing the role of a dark energy component responsible for late-time cosmic acceleration.

The dynamics of the late-time universe are discussed in Ref. [76] to obtain clues useful for resolving the Hubble tension. As the background gravity theory, $f(R)$ gravity is introduced in the metric formalism, and the process of the creation of matter is examined.

In Ref. [77], two kinds of late-time cosmology are studied to explore the deviation from the standard ΛCDM model: (1) a dynamical dark energy originating from the gravitational

 https://doi.org/10.3390/e28020222

field in the expanding universe and (2) a model of the interaction between the dark energy component and matter.

In Ref. [78], a cyclic universe scenario with cosmic contraction and expansion is proposed in the context of the quantum memory matrix cosmology, in which it is assumed that the spacetime is constructed by finite-capacity Hilbert cells.

The authors of Ref. [79] discuss thermodynamics in quasi-de Sitter spacetime, focusing on a specific case where the energy-momentum tensor of spacetime first reproduces a quintessence field asymptotically. A finite domain, in which an additional Cauchy horizon appears, is then analyzed.

We find it very likely that the eight articles summarized above will be useful for future related studies on various aspects of modern cosmology and astrophysics.

Funding: This work was supported by the JSPS KAKENHI—under Grant Numbers 24KF0100 and 25KF0176—and the Competitive Research Funds for Fukushima University Faculty (25RK011).

Conflicts of Interest: The author declares no conflicts of interest.

References

1. Perlmutter, S.; Aldering, G.; Goldhaber, G.; Knop, R.A.; Nugent, P.; Castro, P.G.; Deustua, S.; Fabbro, S.; Goobar, A.; Groom, D.E.; et al. Measurements of Omega and Lambda from 42 High-Redshift Supernovae. *Astrophys. J.* **1999**, *517*, 565. [CrossRef]
2. Riess, A.G.; Filippenko, A.V.; Challis, P.; Clocchiatti, A.; Diercks, A.; Garnavich, P.M.; Gilliland, R.L.; Hogan, C.J.; Jha, S.; Kirshner, R.P.; et al. Observational Evidence from Supernovae for an Accelerating Universe and a Cosmological Constant. *Astron. J.* **1998**, *116*, 1009. [CrossRef]
3. Abbott, T.M.C.; Acevedo, M.; Aguena, M.; Alarcon, A.; Allam, S.; Alves, O.; Amon, A.; Andrade-Oliveira, F.; Annis, J.; Armstrong, P.; et al. The Dark Energy Survey: Cosmology Results with ∼1500 New High-redshift Type Ia Supernovae Using the Full 5 yr Data Set. *Astrophys. J. Lett.* **2024**, *973*, L14. [CrossRef]
4. Spergel, D.N.; Verde, L.; Peiris, H.V.; Komatsu, E.; Nolta, M.R.; Bennett, C.L.; Halpern, M.; Hinshaw, G.; Jarosik, N.; Kogut, A.; et al. First Year Wilkinson Microwave Anisotropy Probe (WMAP) Observations: Determination of Cosmological Parameters. *Astrophys. J. Suppl.* **2003**, *148*, 175. [CrossRef]
5. Spergel, D.N.; Bean, R.; Doré, O.; Nolta, M.R.; Bennett, C.L.; Dunkley, J.; Hinshaw, G.; Jarosik, N.; Komatsu, E.; Page, L.; et al. Wilkinson Microwave Anisotropy Probe (WMAP) three year results: Implications for cosmology. *Astrophys. J. Suppl.* **2007**, *170*, 377. [CrossRef]
6. Komatsu, E.; Dunkley, J.; Nolta, M.R.; Bennett, C.L.; Gold, B.; Hinshaw, G.; Jarosik, N.; Larson, D.; Limon, M.; Page, L.E.; et al. Five-Year Wilkinson Microwave Anisotropy Probe (WMAP) Observations: Cosmological Interpretation. *Astrophys. J. Suppl.* **2009**, *180*, 330. [CrossRef]
7. Komatsu, E.; Smith, K.M.; Dunkley, J.; Bennett, C.L.; Gold, B.; Hinshaw, G.; Jarosik, N.; Larson, D.; Nolta, M.R.; Page, L.; et al. Seven-Year Wilkinson Microwave Anisotropy Probe (WMAP) Observations: Cosmological Interpretation. *Astrophys. J. Suppl.* **2011**, *192*, 18. [CrossRef]
8. Aghanim, N.; Akrami, Y.; Ashdown, M.; Aumont, J.; Baccigalupi, C.; Ballardini, M.; Banday, A.J.; Barreiro, R.B.; Bartolo, N.; Basak, S.; et al. Planck 2018 results. VI. Cosmological parameters. *Astron. Astrophys.* **2020**, *641*, A6, Erratum in *Astron. Astrophys.* **2021**, *652*, C4.
9. Akrami, Y.; Arroja, F.; Ashdown, M.; Aumont, J.; Baccigalupi, C.; Ballardini, M.; Banday, A.J.; Barreiro, R.B.; Bartolo, N.; Basak, S.; et al. Planck 2018 results. X. Constraints on inflation. *Astron. Astrophys.* **2020**, *641*, A10.
10. Louis, T.; La Posta, A.; Atkins, Z.; Jense, H.T.; Abril-Cabezas, I.; Addison, G.E.; Ade, P.A.R.; Aiola, S.; Alford, T.; Alonso, D.; et al. The Atacama Cosmology Telescope: DR6 power spectra, likelihoods and ΛCDM parameters. *JCAP* **2025**, *11*, 062. [CrossRef]
11. Ge, F.; Millea, M.; Camphuis, E.; Daley, C.; Huang, N.; Omori, Y.; Quan, W.; Anderes, E.; Anderson, A.J.; Ansarinejad, B.; et al. Cosmology from CMB lensing and delensed EE power spectra using 2019–2020 SPT-3G polarization data. *Phys. Rev. D* **2025**, *111*, 083534. [CrossRef]
12. Tegmark, M.; Strauss, M.A.; Blanton, M.R.; Abazajian, K.; Dodelson, S.; Sandvik, H.; Wang, X.; Weinberg, D.H.; Zehavi, I.; Bahcall, N.A.; et al. Cosmological parameters from SDSS and WMAP. *Phys. Rev. D* **2004**, *69*, 103501. [CrossRef]
13. Seljak, U.; Makarov, A.; McDonald, P.; Anderson, S.F.; Bahcall, N.A.; Brinkmann, J.; Burles, S.; Cen, R.; Doi, M.; Gunn, J.E.; et al. Cosmological parameter analysis including SDSS Ly-alpha forest and galaxy bias: Constraints on the primordial spectrum of fluctuations, neutrino mass, and dark energy. *Phys. Rev. D* **2005**, *71*, 103515. [CrossRef]

https://doi.org/10.3390/e28020222

14. Tsagas, C.G.; Challinor, A.; Maartens, R. Relativistic cosmology and large-scale structure. *Phys. Rept.* **2008**, *465*, 61–147. [CrossRef]
15. Eisenstein, D.J.; Zehavi, I.; Hogg, D.W.; Scoccimarro, R.; Blanton, M.R.; Nichol, R.C.; Scranton, R.; Seo, H.J.; Tegmark, M.; Zheng, Z.; et al. Detection of the Baryon Acoustic Peak in the Large-Scale Correlation Function of SDSS Luminous Red Galaxies. *Astrophys. J.* **2005**, *633*, 560. [CrossRef]
16. Alam, S.; Aubert, M.; Avila, S.; Balland, C.; Bautista, J.E.; Bershady, M.A.; Bizyaev, D.; Blanton, M.R.; Bolton, A.S.; Bovy, J.; et al. Completed SDSS-IV extended Baryon Oscillation Spectroscopic Survey: Cosmological implications from two decades of spectroscopic surveys at the Apache Point Observatory. *Phys. Rev. D* **2021**, *103*, 083533. [CrossRef]
17. Abdul-Karim, M.; Aguilar, J.; Ahlen, S.; Prieto, C.A.; Alves, O.; Anand, A.; Andrade, U.; Armengaud, E.; Aviles, A.; Bailey, S.; et al. DESI DR2 results. I. Baryon acoustic oscillations from the Lyman alpha forest. *Phys. Rev. D* **2025**, *112*, 083514. [CrossRef]
18. Abdul-Karim, M.; Aguilar, J.; Ahlen, S.; Alam, S.; Allen, L.; Prieto, C.A.; Alves, O.; Anand, A.; Andrade, U.; Armengaud, E.; et al. DESI DR2 results. II. Measurements of baryon acoustic oscillations and cosmological constraints. *Phys. Rev. D* **2025**, *112*, 083515. [CrossRef]
19. Jain, B.; Taylor, A. Cross-correlation Tomography: Measuring Dark Energy Evolution with Weak Lensing. *Phys. Rev. Lett.* **2003**, *91*, 141302. [CrossRef]
20. Munshi, D.; Valageas, P.; Van Waerbeke, L.; Heavens, A. Cosmology with Weak Lensing Surveys. *Phys. Rept.* **2008**, *462*, 67–121. [CrossRef]
21. Troxel, M.A.; Ishak, M. The Intrinsic Alignment of Galaxies and its Impact on Weak Gravitational Lensing in an Era of Precision Cosmology. *Phys. Rept.* **2014**, *558*, 1–59. [CrossRef]
22. Abbott, T.M.C.; Abdalla, F.B.; Alarcon, A.; Aleksić, J.; Allam, S.; Allen, S.; Amara, A.; Annis, J.; Asorey, J.; Avila, S.; et al. Dark Energy Survey year 1 results: Cosmological constraints from galaxy clustering and weak lensing. *Phys. Rev. D* **2018**, *98*, 043526. [CrossRef]
23. Abbott, T.M.C.; Aguena, M.; Alarcon, A.; Allam, S.; Alves, O.; Amon, A.; Andrade-Oliveira, F.; Annis, J.; Avila, S.; Bacon, D.; et al. Dark Energy Survey Year 3 results: Cosmological constraints from galaxy clustering and weak lensing. *Phys. Rev. D* **2022**, *105*, 023520. [CrossRef]
24. Guth, A.H. The Inflationary Universe: A Possible Solution to the Horizon and Flatness Problems. *Phys. Rev. D* **1981**, *23*, 347. [CrossRef]
25. Sato, K. First Order Phase Transition of a Vacuum and Expansion of the Universe. *Mon. Not. Roy. Astron. Soc.* **1981**, 195, 467–479. [CrossRef]
26. Starobinsky, A.A. A New Type of Isotropic Cosmological Models Without Singularity. *Phys. Lett.* **1980**, *91B*, 99. [CrossRef]
27. Linde, A.D. A New Inflationary Universe Scenario: A Possible Solution of the Horizon, Flatness, Homogeneity, Isotropy and Primordial Monopole Problems. *Phys. Lett.* **1982**, *108B*, 389. [CrossRef]
28. Sato, K.; Yokoyama, J. Inflationary cosmology: First 30+ years. *Int. J. Mod. Phys. D* **2015**, 24, 1530025. [CrossRef]
29. Odintsov, S.D.; Oikonomou, V.K.; Giannakoudi, I.; Fronimos, F.P.; Lymperiadou, E.C. Recent Advances in Inflation. *Symmetry* **2023**, 15, 1701. [CrossRef]
30. Kallosh, R.; Linde, A. On the present status of inflationary cosmology. *Gen. Rel. Grav.* **2025**, 57, 135. [CrossRef]
31. Nojiri, S.; Odintsov, S.D. Introduction to modified gravity and gravitational alternative for dark energy. eConf **C0602061**, 06 (2006). *Int. J. Geom. Meth. Mod. Phys.* **2007**, *4*, 115. [CrossRef]
32. Lue, A. The phenomenology of dvali-gabadadze-porrati cosmologies. *Phys. Rept.* **2006**, *423*, 1–48. [CrossRef]
33. Copeland, E.J.; Sami, M.; Tsujikawa, S. Dynamics of dark energy. *Int. J. Mod. Phys. D* **2006**, *15*, 1753–1936. [CrossRef]
34. Fujii, Y.; Maeda, K. *The Scalar-Tensor Theory of Gravitation*; Cambridge University Press: Cambridge, UK, 2007.
35. Padmanabhan, T. Dark energy and gravity. *Gen. Rel. Grav.* **2008**, *40*, 529–564. [CrossRef]
36. Durrer, R.; Maartens, R. Dark Energy and Dark Gravity. *Gen. Rel. Grav.* **2008**, *40*, 301–328. [CrossRef]
37. Alexander, S.; Yunes, N. Chern-Simons Modified General Relativity. *Phys. Rept.* **2009**, *480*, 1–55. [CrossRef]
38. Sotiriou, T.P.; Faraoni, V. $f(R)$ Theories Of Gravity. *Rev. Mod. Phys.* **2010**, *82*, 451–497. [CrossRef]
39. Cai, Y.F.; Saridakis, E.N.; Setare, M.R.; Xia, J.Q. Quintom Cosmology: Theoretical implications and observations. *Phys. Rept.* **2010**, *493*, 1–60. [CrossRef]
40. De Felice, A.; Tsujikawa, S. $f(R)$ theories. *Living Rev. Rel.* **2010**, *13*, 3. [CrossRef]
41. Amendola, L.; Tsujikawa, S. *Dark Energy*; Cambridge University Press: Cambridge, UK, 2010.
42. Faraoni, V.; Capozziello, S. *Beyond Einstein Gravity*; Springer: Dordrecht, The Netherlands, 2010.
43. Nojiri, S.; Odintsov, S.D. Unified cosmic history in modified gravity: From $F(R)$ theory to Lorentz non-invariant models. *Phys. Rept.* **2011**, *505*, 59–144. [CrossRef]
44. Capozziello, S.; De Laurentis, M. Extended Theories of Gravity. *Phys. Rept.* **2011**, *509*, 167–321. [CrossRef]
45. Clifton, T.; Ferreira, P.G.; Padilla, A.; Skordis, C. Modified Gravity and Cosmology. *Phys. Rept.* **2012**, *513*, 1–189. [CrossRef]

https://doi.org/10.3390/e28020222

46. Weinberg, D.H.; Mortonson, M.J.; Eisenstein, D.J.; Hirata, C.; Riess, A.G.; Rozo, E. Observational Probes of Cosmic Acceleration. *Phys. Rept.* **2013**, *530*, 87–255. [CrossRef]
47. Bamba, K.; Capozziello, S.; Nojiri, S.; Odintsov, S.D. Dark energy cosmology: The equivalent description via different theoretical models and cosmography tests. *Astrophys. Space Sci.* **2012**, *342*, 155–228. [CrossRef]
48. Padmanabhan, T.; Kothawala, D. Lanczos-Lovelock models of gravity. *Phys. Rept.* **2013**, *531*, 115–171. [CrossRef]
49. Will, C.M. The Confrontation between General Relativity and Experiment. *Living Rev. Rel.* **2014**, *17*, 4. [CrossRef] [PubMed]
50. Joyce, A.; Jain, B.; Khoury, J.; Trodden, M. Beyond the Cosmological Standard Model. *Phys. Rept.* **2015**, *568*, 1–98. [CrossRef]
51. Bamba, K.; Odintsov, S.D. Inflationary cosmology in modified gravity theories. *Symmetry* **2015**, *7*, 220–240. [CrossRef]
52. Cai, Y.F.; Capozziello, S.; De Laurentis, M.; Saridakis, E.N. $f(T)$ teleparallel gravity and cosmology. *Rept. Prog. Phys.* **2016**, *79*, 106901. [CrossRef]
53. Wang, S.; Wang, Y.; Li, M. Holographic Dark Energy. *Phys. Rept.* **2017**, *696*, 1–57. [CrossRef]
54. Nojiri, S.; Odintsov, S.D.; Oikonomou, V.K. Modified Gravity Theories on a Nutshell: Inflation, Bounce and Late-time Evolution. *Phys. Rept.* **2017**, *692*, 1–104. [CrossRef]
55. Beltran Jimenez, J.; Heisenberg, L.; Olmo, G.J.; Rubiera-Garcia, D. Born–Infeld inspired modifications of gravity. *Phys. Rept.* **2018**, *727*, 1–129. [CrossRef]
56. Bahamonde, S.; Böhmer, C.G.; Carloni, S.; Copeland, E.J.; Fang, W.; Tamanini, N. Dynamical systems applied to cosmology: Dark energy and modified gravity. *Phys. Rept.* **2018**, *775–777*, 1–122. [CrossRef]
57. Adami, H.; Setare, M.R.; Sisman, T.C.; Tekin, B. Conserved Charges in Extended Theories of Gravity. *Phys. Rept.* **2019**, *834*, 1. [CrossRef]
58. Heisenberg, L. A systematic approach to generalisations of General Relativity and their cosmological implications. *Phys. Rept.* **2019**, *796*, 1–113. [CrossRef]
59. Langlois, D. Dark energy and modified gravity in degenerate higher-order scalar–tensor (DHOST) theories: A review. *Int. J. Mod. Phys. D* **2019**, *28*, 1942006. [CrossRef]
60. Kobayashi, T. Horndeski theory and beyond: A review. *Rept. Prog. Phys.* **2019**, *82*, 086901. [CrossRef]
61. Frusciante, N.; Perenon, L. Effective field theory of dark energy: A review. *Phys. Rept.* **2020**, *857*, 1–63. [CrossRef]
62. Olmo, G.J.; Rubiera-Garcia, D.; Wojnar, A. Stellar structure models in modified theories of gravity: Lessons and challenges. *Phys. Rept.* **2020**, *876*, 1–75. [CrossRef]
63. Saridakis, E.N.; Lazkoz, R.; Salzano, V.; Moniz, P.V.; Capozziello, S.; Jimenez, J.B.; De Laurentis, M.; Olmo, G.J. *Modified Gravity and Cosmology*; Springer: Cham, Switzerland, 2021.
64. Faraoni, V.; Giusti, A.; Fahim, B.H. Spherical inhomogeneous solutions of Einstein and scalar–tensor gravity: A map of the land. *Phys. Rept.* **2021**, *925*, 1–58. [CrossRef]
65. Bamba, K. Review on Dark Energy Problem and Modified Gravity Theories. *LHEP* **2022**, *2022*, 352. [CrossRef]
66. Bahamonde, S.; Dialektopoulos, K.F.; Escamilla-Rivera, C.; Farrugia, G.; Gakis, V.; Hendry, M.; Hohmann, M.; Levi Said, J.; Mifsud, J.; Di Valentino, E. Teleparallel gravity: From theory to cosmology. *Rept. Prog. Phys.* **2023**, 86, 026901. [CrossRef]
67. Arai, S.; Aoki, K.; Chinone, Y.; Kimura, R.; Kobayashi, T.; Miyatake, H.; Yamauchi, D.; Yokoyama, S.; Akitsu, K.; Hiramatsu, T.; et al. Cosmological gravity probes: Connecting recent theoretical developments to forthcoming observations. *PTEP* **2023**, *2023*, 072E01. [CrossRef]
68. de Haro, J.; Nojiri, S.; Odintsov, S.D.; Oikonomou, V.K.; Pan, S. Finite-time cosmological singularities and the possible fate of the Universe. *Phys. Rept.* **2023**, *1034*, 1–114. [CrossRef]
69. Heisenberg, L. Review on f(Q) gravity. *Phys. Rept.* **2024**, *1066*, 1–78. [CrossRef]
70. Avsajanishvili, O.; Chitov, G.Y.; Kahniashvili, T.; Mandal, S.; Samushia, L. Observational Constraints on Dynamical Dark Energy Models. *Universe* **2024**, 10, 122. [CrossRef]
71. Yousaf, Z.; Bamba, K.; Bhatti, M.Z.; Farwa, U. Quasi-static evolution of axially and reflection symmetric large-scale configuration. *Int. J. Geom. Meth. Mod. Phys.* **2024**, *21*, 2430005. [CrossRef]
72. Tan, X.; Wang, C.; Yang, S.Z. Corrections to the Bekenstein-Hawking Entropy of the HNUTKN Black Hole Due to Lorentz-Breaking Fermionic Einstein-Aether Theory. *Entropy* **2024**, *26*, 326. [CrossRef] [PubMed]
73. Herrera, L.; Witten, L. The Hyperbolically Symmetric Black Hole. *Entropy* **2025**, *27*, 831. [CrossRef] [PubMed]
74. Wang, C.; Zhang, J.; Yang, S.Z. Entropy Modifications of Charged Accelerating Anti-de Sitter Black Hole. *Entropy* **2025**, *27*, 900. [CrossRef] [PubMed]
75. Wang, C. Correction to Temperature and Bekenstein–Hawking Entropy of Kiselev Black Hole Surrounded by Quintessence. *Entropy* **2025**, *27*, 1135. [CrossRef] [PubMed]
76. Montani, G.; Carlevaro, N.; De Angelis, M. Modified Gravity in the Presence of Matter Creation: Scenario for the Late Universe. *Entropy* **2024**, *26*, 662. [CrossRef]
77. Montani, G.; Fazzari, E.; Carlevaro, N.; Dainotti, M.G. Two Dynamical Scenarios for Binned Master Sample Interpretation. *Entropy* **2025**, *27*, 895. [CrossRef]

https://doi.org/10.3390/e28020222

78. Neukart, F.; Marx, E.; Vinokur, V. Counting Cosmic Cycles: Past Big Crunches, Future Recurrence Limits, and the Age of the Quantum Memory Matrix Universe. *Entropy* **2025**, *27*, 1043. [CrossRef] [PubMed]
79. Luongo, O.; Azizinia, M.; Boshkayev, K. Entropy of a Quasi-de Sitter Spacetime and the Role of Specific Heat. *Entropy* **2026**, *28*, 43. [CrossRef] [PubMed]

https://doi.org/10.3390/e28020222

entropy

Article

Corrections to the Bekenstein–Hawking Entropy of the HNUTKN Black Hole Due to Lorentz-Breaking Fermionic Einstein–Aether Theory

Xia Tan [1], Cong Wang [1,*] and Shu-Zheng Yang [2]

[1] College of Physics and Electronic Engineering, Qilu Normal University, Jinan 250300, China; 20152329@qlnu.edu.cn
[2] College of Physics and Astronomy, China West Normal University, Nanchong 637002, China; szyang@cwnu.edu.cn
* Correspondence: wangcong@qlnu.edu.cn

Abstract: A hot NUT–Kerr–Newman black hole is a general stationary axisymmetric black hole. In this black hole spacetime, the dynamical equations of fermions at the horizon are modified by considering Lorentz breaking. The corrections to the Hawking temperature and Bekenstein–Hawking entropy at the horizon of the black hole are studied in depth. Based on the semiclassical theory correction, the Bekenstein–Hawking entropy of this black hole is quantum-corrected by considering the perturbation effect of the Planck constant $\hbar$. The latter part of this paper presents a detailed discussion of the obtained results and their physical implications.

Keywords: hot NUT–Kerr–Newman black hole; Lorentz breaking; black hole entropy

1. Introduction

It is well known that de Sitter spacetime can be thought of as being hot, since an observer moving along a geodesic will detect an isotropic background of thermal radiation with a temperature T. NUT–Kerr–Newman–de Sitter spacetimes are asymptotically de Sitter instead of being asymptotically flat. This is a more general axisymmetric background. We call the spacetime concerned here the NUT (Newman–Unti–Tamburino)–Kerr–Newman spacetime, since the de Sitter spacetime has been interpreted as being hot [1]. For brevity, we refer to hot NUT (Newman–Unti–Tamburino)–Kerr–Newman as HNUTKN. The purpose of the present paper is to study the corrections and physical implications of entropy for the hot NUT–Kerr–Newman (HNUTKN) black hole. Reference [2] investigates the Hawking radiation of Dirac particles in the spacetime, obtaining a pure thermal radiation spectrum. However, the actual quantum tunneling radiation rate of the black hole is related to the Bekenstein–Hawking entropy change, and the thermal radiation spectrum of the black hole should not be a blackbody spectrum. Research on the correction of quantum tunneling radiation in HNUTKN black holes could lead to new expressions for the quantum tunneling rate, Hawking temperature, and Bekenstein–Hawking entropy of this black hole. Only by considering the real quantum tunneling radiation can we provide some explanation for the puzzle of the black hole information loss, thus making the research more physically meaningful. The research method for the entropy of HNUTKN black holes and the series of results obtained in this paper include corresponding results for the NUT–Kerr–Newman spacetime, NUT–Kerr–Newman–de Sitter spacetime, Kerr–Newman spacetime, and Kerr spacetime. Therefore, it is necessary to conduct in-depth research on the correction of entropy for HNUTKN black holes.

The first and second laws of black hole thermodynamics are both related to black hole entropy, which in turn is associated with black hole radiation, indicating a close connection to black hole information. Recent research results have shown that the thermal radiation spectrum of a black hole is no longer strictly a blackbody spectrum but contains information that escapes with the radiation, thereby implying that no information is lost during the process of black hole thermal radiation [3–5]. Unitarity is one of the fundamental principles of quantum theory, and information conservation is necessary to ensure the unitarity and probability conservation of quantum theory. As a black hole radiates, a portion of the information is carried away beyond the black hole, while another portion remains as remnants, called "ashes", due to the black hole ceasing to emit particles towards the end of its radiation. The mass of these remnants, denoted as M_{res}, is equal to the minimum mass of the black hole M_{min}. The specific heat capacity of a black hole is related to its Hawking temperature and entropy. Recent research has indicated that the generalized uncertainty principle (GUP) can lead to the formation of black hole remnants [6]. Different black holes may possess distinct remnants. The existence of black hole remnants indicates that black holes do not evaporate completely, as predicted by classical theory. In fact, there are still many unresolved issues regarding the remnants and the information paradox.

The calculation methods of black hole entropy include the brick wall model and the membrane model. 't Hooft proposed the brick wall model in 1985 to explain black hole entropy and suggested that the entropy of a quantum gas (Hawking radiation) in thermal equilibrium with the black hole outside is equal to the black hole's entropy [7]. Following this calculation method, Zhao et al. improved 't Hooft's brick wall model to obtain the membrane model and used this model to study black hole entropy [8–10]. According to the viewpoint of the membrane model, black hole entropy is actually the entropy of two-dimensional membranes on the black hole event horizon, which are curved surfaces intersecting the three-dimensional space at the same time. This entropy is contributed by the two-dimensional quantum gas on the event horizon, requiring the inclusion of a thin layer outside the black hole event horizon with a thickness ε and a distance d from the event horizon when calculating black hole entropy. By studying the entropy of the quantum gas in this thin layer and taking the limits $\varepsilon \to 0$ and $d \to 0$, the black hole entropy can be obtained. Research results show that the Bekenstein–Hawking entropy of a black hole is proportional to its event horizon area A, i.e., $S_{BH} \propto A/4$. For a HNUTKN black hole, according to the first law of black hole thermodynamics, its general expression in differential form is $\mathrm{d}M = T\mathrm{d}S_{BH} + \Omega \mathrm{d}J + Q\mathrm{d}V$, where M represents the Bekenstein–Smarr mass. By studying the quantum tunneling radiation of a black hole, the Hawking temperature at the event horizon can be determined. Since the relationship between the Hawking temperature T and the surface gravity κ of the event horizon is given by $T = \kappa/2\pi$, the study of the quantum tunneling radiation characteristics and the first law of black hole thermodynamics can be employed to investigate the Bekenstein–Hawking entropy of black holes. However, recent research on quantum gravity theories suggests that Lorentz relations need appropriate modifications in high-energy regimes. Before a theory of dispersion relations in the high-energy regime is established, it is generally believed that the order of magnitude of the correction terms for Lorentz dispersion relations is at the Planck scale [11–15]. The appearance of such correction terms has prompted investigations into the effects of Lorentz-breaking theories in high-energy physics on curved spacetime backgrounds and the modifications of the quantum tunneling radiation characteristics under specific curved spacetime backgrounds. In 2009, Horava proposed the Horava–Lifshitz gravity theory, which breaks the Lorentz symmetry by not having temporal and spatial symmetries in high-energy situations. Another Lorentz-breaking gravity theory is the Einstein–aether theory. The Horava–Lifshitz gravity theory and the Einstein–aether theory are consistent in low-energy situations. The effects of Lorentz-breaking on steady-state axisymmetric spacetime backgrounds are reflected in the spacetime metric [16]. Considering Lorentz breaking, modifications to the dynamical equations of bosons and fermions and in-depth studies of the quantum tunneling radiation characteristics of different black holes in specific curved spacetime backgrounds are topics worth exploring [17–19]. Through such research, we can delve into the entropy of black holes and gain a profound understanding of the physical significance of black hole entropy and information.

Section 2 below pertains to the modified forms of the dynamical equations of fermions with Lorentz breaking. Section 3 of this paper focuses on the research methods and physical significance of the obtained results on HNUTKN black hole entropy. Section 4 of this paper is a discussion of the research results.

2. Lorentz Breaking and the Modified Dynamical Equation of the Spinor Field

When considering the modification of the action of the spinor field with Lorentz-breaking fermionic Einstein–aether terms in a curved spacetime, one needs to consider the aether-like field vector u^μ, the Chiral correction term, and the Carroll–Field–Jackiw (CFJ) correction term. After Carroll et al. proposed the CFJ correction term in 1990, research on modifications to Lorentz breaking has been carried out [20–22], and this aspect of research still has practical significance today. It is necessary to study the influence of Lorentz breaking on black hole quantum tunneling radiation and black hole entropy in different curved spacetimes. Before modifying the black hole entropy, one should first consider the effect of Lorentz breaking on the action of the spinor field. Introducing a Lorentz-breaking correction term in a curved spacetime can express the action of the spinor field as [17,23,24]

$$\begin{aligned}S_F = \int d^4x\sqrt{-g}\bar{\psi}_{\alpha_1\ldots\alpha_k}\Big\{ i\gamma^\mu\left(\partial_\mu - \frac{ie}{\hbar}A_\mu + i\Omega_\mu\right)\left[1 - \frac{a'\hbar^2}{m^2}\gamma^\alpha\left(\partial_\alpha - \frac{ie}{\hbar}A_\alpha + i\Omega_\alpha\right)\gamma^\beta\left(\partial_\beta - \frac{ie}{\hbar}A_\beta + i\Omega_\beta\right)\right] \\ + \frac{\lambda}{m}[u^\alpha(\partial_\alpha - ieA_\alpha + i\Omega_\alpha)]^2 + \frac{b'}{\hbar m}\gamma^5 - \frac{m}{\hbar}\Big\}\psi_{\alpha_1\ldots\alpha_k}\end{aligned} \tag{1}$$

Here, a', b', and λ are all small dimensionless real numbers. b' is the coefficient for Chiral correction terms; a' is the coefficient for CFJ correction terms. γ^μ, γ^α, and γ^β are gamma matrices in a curved spacetime; u^α and u^β are aether-like field vectors in a curved spacetime; and u^μ must satisfy $u^\mu u_\mu =$ const. γ^5 is the gamma matrix corresponding to the Chiral correction term. Since γ^5 appears in S_F, it is required that γ^5 is also a Hermitian matrix. $\Omega_\mu, \Omega_\alpha$, and Ω_β are the spin connections. $\Omega_\mu = \frac{i}{2}\Gamma_\mu^{\alpha\beta}\Pi_{\alpha\beta}, \Pi_{\alpha\beta} = \frac{i}{4}\left[\gamma^\alpha, \gamma^\beta\right]$, where $\Gamma_\mu^{\alpha\beta}$ is the connection of Riemann space and is related to the metric tensor of the curved spacetime. $A_\mu(A_\alpha, A_\beta)$ is the electromagnetic potential generated by a black hole charge. $\psi_{\alpha_1\ldots\alpha_k}$ represents the wave function of fermions. $\bar{\psi}_{\alpha_1\ldots\alpha_k}$ is the Hermitian conjugate of $\psi_{\alpha_1\ldots\alpha_k}$. The relationship between the gamma matrices in Equation (1) and the metric tensor in a curved spacetime is as follows:

$$\gamma^\mu\gamma^\nu + \gamma^\nu\gamma^\mu = 2g^{\mu\nu}\mathrm{I} \tag{2}$$

$$\gamma^\mu\gamma^5 + \gamma^5\gamma^\mu = 0 \tag{3}$$

where I is the identity matrix. According to the variational principle, we can derive the spinor field equations in a curved spacetime from Equation (1). Substituting S_F into the following equation

$$\delta S_F = 0, \tag{4}$$

we obtain

$$\begin{aligned}\delta\Big\{\bar{\psi}_{\alpha_1\ldots\alpha_k}\Big[i\gamma^\mu\left(\partial_\mu - \frac{ie}{\hbar}A_\mu + i\Omega_\mu\right)\left[1 - \frac{a'\hbar^2}{m^2}\gamma^\alpha\gamma^\beta\left(\partial_\alpha - \frac{ie}{\hbar}A_\alpha + i\Omega_\alpha\right)\left(\partial_\beta - \frac{ie}{\hbar}A_\beta + i\Omega_\beta\right)\right] \\ + \frac{\hbar\lambda}{m} + \frac{b'}{\hbar m}\gamma^5 - \frac{m}{\hbar}\Big]\Big\}\psi_{\alpha_1\ldots\alpha_k} = 0\end{aligned} \tag{5}$$

The equation shows that

$$\delta\left[\left(\frac{b'}{\hbar m}\gamma^5 - \frac{m}{\hbar}\right)\psi_{\alpha_1\ldots\alpha_k}\right] = \left(\frac{b'}{\hbar m}\gamma^5 - \frac{m}{\hbar}\right)\partial_\mu\psi_{\alpha_1\ldots\alpha_k}\delta x^\mu \tag{6}$$

$$\delta\left[u^\alpha\left(\partial_\alpha - \frac{ie}{\hbar}A_\alpha + i\Omega_\alpha\right)\right]^2\psi_{\alpha_1\ldots\alpha_k} = u^\alpha u^\beta\left(\partial_\alpha - \frac{ie}{\hbar}A_\alpha + i\Omega_\alpha\right)\left(\partial_\beta - \frac{ie}{\hbar}A_\beta + i\Omega_\beta\right)\partial_\mu\psi_{\alpha_1\ldots\alpha_k}\delta x^\mu \tag{7}$$

Noting that $\delta\bar{\psi}_{\alpha_1......\alpha_k} = \partial\mu\bar{\psi}_{\alpha_1......\alpha_k}\delta x^\mu$ and $\delta\psi_{\alpha_1......\alpha_k} = \partial\mu\psi_{\alpha_1......\alpha_k}\delta x^\mu$, the modified spinor field equation can be obtained as follows from Equations (5)–(7):

$$\begin{aligned} &i\gamma^\mu\left(\partial_\mu - \frac{i\varphi}{\hbar}A_\mu + \Omega_\mu\right)\left[1 - \frac{a'\hbar^2}{m^2}\gamma^\alpha\gamma^\beta\left(\partial_\alpha - \frac{ie}{\hbar}A_\alpha + i\Omega_\alpha\right)\left(\partial_\beta - \frac{ie}{\hbar}A_\beta + i\Omega_\beta\right)\right]\psi_{\alpha_1...\alpha_k} \\ &= \left[\frac{m}{\hbar} - \frac{b'}{\hbar m}\gamma^5 - \frac{\hbar\lambda}{m}u^\alpha u^\beta(\partial_\alpha - ieA_\alpha + i\Omega_\alpha)(\partial_\beta - ieA_\beta + i\Omega_\beta)\right]\psi_{\alpha_1...\alpha_k}. \end{aligned} \tag{8}$$

The equation for $\bar{\psi}_{\alpha_1......\alpha_k}$ is the Hermitian conjugate of Equation (8). Equation (8) is a modified form of the fermion dynamical equation. The expression $\psi_{\alpha_1......\alpha_k}$ in this equation represents the wave function of an arbitrary spin fermion. This equation applies to any spin fermion. For spin-1/2 fermions, the wave function $\psi_{\alpha_1...\alpha_k}$ in Equation (8) can be expressed using the WKB theory:

$$\psi_{\frac{1}{2}} = \begin{pmatrix} A \\ B \end{pmatrix} e^{\frac{i}{\hbar}S_a} \tag{9}$$

Here, S_a is the action for spin-1/2 fermions, which refers to Dirac particles in a curved spacetime. In this case, Equation (8) is known as the modified form of the Dirac equation. For spin-3/2 fermions, the wave function in Equation (8) is denoted as $\psi_{\alpha_1......\alpha_k} \to \psi_\nu$, and the equation is then referred to as the modified form of the Rarita–Schwinger equation. In studying the solutions of Equation (8) in this scenario, the following conditions must be satisfied:

$$\gamma^\nu\psi_\nu = 0. \tag{10}$$

In general, for fermions of any spin, the wave function in Equation (8) should be substituted as $\psi_{\alpha_1......\alpha_k}$, and the corresponding additional condition for Equation (8) is

$$\gamma^\mu\psi_{\mu\alpha_2......\alpha_k} = D_\mu\psi^\mu{}_{\alpha_2......\alpha_k} = \psi^\mu{}_{\mu\alpha_3......\alpha_k} = 0. \tag{11}$$

When $k = 0$, Equation (8) reduces to the modified form of the Dirac equation, in which case there are no additional conditions. When $k = 1$, Equation (11) reverts back to Equation (10), and, in this scenario, Equation (8) represents the modified form of the Rarita–Schwinger equation. When studying the dynamical characteristics of fermions with different spins in a specific black hole spacetime, the methods and processes for solving are not the same. When analyzing the dynamical equations at the black hole horizon using semiclassical theory, condition (11) does not need to be considered, and the calculation of the quantum tunneling rates will yield the same results. To illustrate the application of Equation (8) more clearly, let us consider spin-1/2 fermions, i.e., Dirac particles. Substituting Equation (9) into Equation (8), we obtain the equation that the particle action S_a must satisfy:

$$\begin{aligned} &\left\{-i\gamma^\mu(\partial_\mu S_a - eA_\mu)\left[1 + \frac{a'}{m^2}g^{\alpha\beta}(\partial_\alpha S_a - eA_\alpha)(\partial_\beta S_a - eA_\beta)\right]\right. \\ &\left. - m + \frac{b'}{m}\gamma_0^5 + \frac{\lambda}{m}u^\alpha u^\beta(\partial_\alpha S_a - eA_\alpha)(\partial_\beta S_a - eA_\beta)\right\}\begin{pmatrix} A \\ B \end{pmatrix} = 0 \end{aligned} \tag{12}$$

In this equation, several small terms, such as $\hbar\Omega_\mu$, $\hbar\Omega_\alpha$, and $\hbar\Omega_\beta$, have been neglected. Therefore, Equation (12) represents a semiclassical, Lorentz-breaking modified dynamical equation for spin-1/2 fermions, also known as the spin-1/2 spinor field equation. The matrices $g^{\alpha\beta}$ and $\gamma^\alpha\gamma^\beta$ in this matrix equation must satisfy Equation (2). γ_0^5 is closely related to γ^5, requiring γ^5 to satisfy Equation (3) and γ_0^5 to be a real number. Thus, it is required that $\gamma^5 = \gamma_0^5 I$, where I denotes the identity matrix. For the matrix Equation (12) to have solutions, the corresponding determinant of the matrices in the equation must be equal to zero, i.e.,

$$\begin{aligned}&\gamma^{\mu}(\partial_{\mu}S_a - eA_{\mu})\left[1 + \frac{a'}{m^2}g^{\alpha\beta}(\partial_{\alpha}S_a - eA_{\alpha})(\partial_{\beta}S_a - eA_{\beta})\right]\\&= \frac{\lambda}{m}u^{\alpha}u^{\beta}(\partial_{\alpha}S_a - eA_{\alpha})(\partial_{\beta}S_a - eA_{\beta}) + \frac{b'}{m}\gamma_0^5 - m = 0\end{aligned} \tag{13}$$

By multiplying both sides of this equation by $\gamma^{\nu}(\partial_{\nu}S - eA_{\nu})$ and neglecting higher-order small terms, and using Equation (2), we can obtain

$$g^{\mu\nu}(1 + 2a')(\partial_{\mu}S_a - eA_{\mu})(\partial_{\nu}S_a - eA_{\nu}) + 2\lambda u^{\mu}u^{\nu}(\partial_{\mu}S_a - eA_{\mu})(\partial_{\nu}S_a - eA_{\nu}) + 2b'\gamma_0^5 - m^2 = 0 \tag{14}$$

This equation describes the spinor field equation for spin-1/2 fermions, incorporating terms such as the CFJ correction term, aether-like correction term, and Chiral correction term. Equation (14) presents the modified dynamical equations for spin-1/2 fermions utilizing the particle action S_a, with three correction terms included. Equation (14) is applicable for general static, stationary, and non-stationary black hole spacetimes. Therefore, Equation (14) represents a semiclassical spinor field equation, which has been modified to incorporate Lorentz-breaking effects. It should be noted that for fermions of any spin, there are different matrices corresponding to the wave functions and different additional conditions. The modified form of the fermion dynamical equation is represented by the action S_a, as shown in Equation (14).

3. Research on the Entropy of HNUTKN Black Holes

The line element of the HNUTKN black hole spacetime in Boyer–Lindquist coordinates is expressed as

$$\mathrm{d}S^2 = \rho^2\left(\frac{\mathrm{d}r^2}{\Delta_r} + \frac{1}{\Delta_\theta}\mathrm{d}\theta^2\right) + \frac{\sin^2\theta}{\rho^2 J^2}\Delta_\theta\left[a\mathrm{d}t - \left(r^2 + a^2\right)\mathrm{d}\phi\right]^2 - \frac{\Delta_r}{\rho^2 J^2}\left[\mathrm{d}t - \left(a - \frac{(n - a\cos\theta)^2}{a}\right)\mathrm{d}\phi\right]^2 \tag{15}$$

Here,

$$\begin{aligned}\rho^2 &= r^2 + (n - a\cos\theta)^2 = r^2 + n^2 - 2an\cos\theta + a^2\cos^2\theta\\ \Delta_r &= \left(r^2 + a^2 - n^2\right)\left(1 - \frac{1}{3}\Lambda r^2\right) - 2Mr + Q^2\\ \Delta_\theta &= 1 + \frac{1}{3}\Lambda(n - a\cos\theta)^2\\ J &= 1 + \frac{1}{3}\Lambda\left(a^2 - n^2\right)\end{aligned} \tag{16}$$

Here, Λ is the cosmological constant, M is the black hole mass, Q is the black hole charge, the NUT parameter is n, also known as the gravitational magnetic type mass, and the angular momentum parameter is a. Due to the electromagnetic charge Q of the black hole, the electromagnetic potentials A_μ, A_t, and A_ϕ are expressed as

$$A_\mu = \left(-\frac{Qr}{\rho^2 J}, 0, 0, \frac{Qra\sin^2\theta}{\rho^2 J}\right) \tag{17}$$

According to Equations (15) and (16), the determinant corresponding to the metric $g_{\mu\nu}$ is obtained as

$$g = |g_{\mu\nu}| = -\frac{\sin^2\theta}{J^4}\left\{r^2 + a^2 - \left[a^2 - (n - a\cos\theta)^2\right]^2\right\} \tag{18}$$

Using Equation (18) and the algebraic cofactor $(-1)^{\mu+\nu}\Delta^{\mu\nu}$ of $g_{\mu\nu}$, the non-zero components of the contravariant metric tensor corresponding to the line element (15) can be, respectively, calculated as

$$\begin{aligned} g^{tt} &= -\frac{J^2}{\rho^2}\left\{\frac{\left(r^2+a^2\right)^2}{\Delta_r} - \frac{1}{a^2\Delta_\theta \sin^2\theta}\left[a^2-(n-a\cos\theta)^2\right]^2\right\} \\ g^{rr} &= \frac{\Delta_r}{\rho^2} \\ g^{\theta\theta} &= \frac{\Delta_\theta}{\rho^2} \\ g^{\phi\phi} &= \frac{J^2}{\rho^2}\left(\frac{1}{\Delta_\theta\sin^2\theta} - \frac{a^2}{\Delta_r}\right) \\ g^{t\phi} &= \frac{J^2}{\rho^2}\left[\frac{a}{\Delta_r}\left(r^2+a^2\right) - \frac{1}{\Delta_\theta a\sin^2\theta}\left(a^2-(n-a\cos\theta)^2\right)\right] \end{aligned} \tag{19}$$

The Bekenstein–Hawking entropy of the HNUTKN black hole is related to the Hawking temperature at the black hole horizon. Therefore, it is necessary to determine the location of the black hole horizon. According to the null supersurface equation at the black hole horizon, it can be determined that

$$g^{\mu\nu}\frac{\partial F}{\partial x^\mu}\frac{\partial F}{\partial x^\nu} = 0 \tag{20}$$

The normal vector of the hypersurface $F(x^\mu) = 0$ is n_μ, where $n_\mu = \frac{\partial F}{\partial x^\mu}$. The definition of the null hypersurface is

$$n^\mu n_\mu = 0 \tag{21}$$

We have $g^{\mu\nu}n_\mu n_\mu = 0$, so we have the null supersurface Equation (20). Substituting Equation (19) into Equation (20) and taking into account the stationary axisymmetric nature of the HNUTKN black hole, we obtain the equation satisfied by the event horizon r_H or the cosmological horizon r_c of this black hole as

$$g^{rr} = \frac{\Delta_r}{\rho^2} = 0 \rightarrow \Delta_r \rightarrow 0 \tag{22}$$

Specifically,

$$\left(r^2+a^2-n^2\right)\left(1-\frac{1}{3}\Lambda r^2\right) - 2Mr + Q^2 = 0 \tag{23}$$

In order to solve Equation (14), it is necessary to correctly choose the aether-like vector field u^μ according to the line element (15) as follows:

$$\begin{aligned} u^t &= \frac{c_t}{\sqrt{g_{tt}}}, u^t u_t = u^t u^t g_{tt} = c_t^2 \\ u^r &= \frac{c_r}{\sqrt{g_{rr}}}, u^r u_r = u^r u^r g_{rr} = c_r^2 \\ u^\theta &= \frac{c_\theta}{\sqrt{g_{\theta\theta}}}, u^\theta u_\theta = u^\theta u^\theta g_{\theta\theta} = c_\theta^2 \\ u^\phi &= \frac{c_\phi}{\sqrt{g_{\phi\phi}}}, u^\phi u_\phi = u^\phi u^\phi g_{\phi\phi} = c_\phi^2 \end{aligned} \tag{24}$$

Clearly, u^μ satisfies $u^\mu u_\mu =$ const. In order to determine γ_0^5 in Equation (14), we must choose γ^μ satisfying Equation (2) and the Hermitian matrix γ^5 satisfying Equation (3). Therefore, we choose the frame field e_{I}^μ and determine the gamma matrix γ^μ as follows:

$$\gamma^\mu = e_{\mathrm{I}}^\mu \gamma^{\mathrm{I}} \tag{25}$$

As a result, we obtain

$$\gamma^{\mu}\gamma^{\nu}+\gamma^{\nu}\gamma^{\mu}=e^{\mu}_{I}\gamma^{I}e^{\nu}_{J}\gamma^{J}+e^{\nu}_{J}\gamma^{J}e^{\mu}_{I}\gamma^{I}=2g^{\mu\nu}\mathrm{I} \tag{26}$$

where $e^{\mu}_{I}e^{\nu}_{J}\eta^{IJ}=g^{\mu\nu}$, $e^{\nu}_{J}e^{\mu}_{I}\eta^{JI}=g^{\nu\mu}=g^{\mu\nu}$. Clearly, γ^{μ} satisfies condition (2). Here, the values of I and J are $(0,1,2,3)$, corresponding to the coordinates (t,r,θ,ϕ). We choose γ^5 as follows:

$$\begin{aligned}\gamma^5&=\frac{i\gamma^5_0}{4!}\varepsilon_{\mu\nu k\lambda}\gamma^{\mu}\gamma^{\nu}\gamma^{k}\gamma^{\lambda}\\&=\frac{i\gamma^5_0}{4!}\varepsilon_{IJKL}e^{I}_{\mu}e^{J}_{\nu}e^{I}_{k}e^{J}_{\lambda}e^{\mu}_{I'}\gamma^{I'}e^{\nu}_{I}\gamma^{J'}e^{k}_{k'}\gamma^{k'}e^{\lambda}_{L'}\gamma^{L'}\\&=\frac{i\gamma^5_0}{4!}\varepsilon_{IJKL}\gamma^{I}\gamma^{5}\gamma^{k}\gamma^{L}\\&=i\gamma^5_0\gamma^0\gamma^1\gamma^2\gamma^3\\&=\gamma^5_0\begin{pmatrix}0&\mathrm{I}\\\mathrm{I}&0\end{pmatrix}\end{aligned} \tag{27}$$

Here, γ^5_0 is a real number. Clearly, γ^5 is a Hermitian matrix. ε_{IJKL} is the Levi–Civita symbol, and $\varepsilon_{0123}=1$. Therefore, from Equations (25) and (27), we can obtain

$$e^{\mu}_{I}\gamma^{I}\gamma^{5}+\gamma^{5}e^{\mu}_{I}\gamma^{I}=0 \tag{28}$$

$$\gamma^{\mu}\gamma^{5}+\gamma^{5}\gamma^{\mu}=e^{\mu}_{I}\gamma^{I}\gamma^{5}+\gamma^{5}e^{\mu}_{I}\gamma^{I}=e^{\mu}_{I}\left(\gamma^{I}\gamma^{5}+\gamma^{5}\gamma^{I}\right)=0 \tag{29}$$

From these two equations, it can be seen that choosing γ^5 satisfies condition (3). The correct selection of γ^5, γ^{μ}, and u^{μ} ensures the scientific validity of Equation (14). Substituting Equations (17), (18), (24), and (27) into Equation (14), we obtain the dynamic equation for spin-1/2 fermions in the HNUTKN black hole spacetime as follows:

$$\begin{aligned}&\left(1+2a'+2\lambda c_t^2\right)J^2\left\{\left[\frac{(a^2-(n-a\cos\theta)^2)^2}{a^2\Delta_\theta\sin^2\theta}-\frac{(r^2+a^2)^2}{\Delta_r}\right]\left[\left(\frac{\partial S_a}{\partial t}\right)^2\rho^2+\frac{2eQr}{J}\frac{\partial S_a}{\partial_t}+\frac{e^2Q^2r^2}{\rho^2J^2}\right]\right.\\&+2\left[\frac{a(r^2+a^2)}{\Delta_r}-\frac{a^2-(n-a\cos\theta)^2}{\Delta_\theta a\sin^2\theta}\right]\left[\left(\frac{\partial S_a}{\partial t}\right)\left(\frac{\partial S_a}{\partial_\phi}\right)\rho^2-\frac{eQra\sin^2\theta}{J}\frac{\partial S_a}{\partial_t}+\frac{eQr}{J}\frac{\partial S_a}{\partial_\phi}-\frac{e^2Q^2r^2a\sin^2\theta}{\rho^2J^2}\right]\\&\left.+\left(\frac{1}{\Delta_\theta\sin^2\theta}-\frac{a^2}{\Delta_r}\right)\left[\left(\frac{\partial S_a}{\partial\phi}\right)^2\rho^2-\frac{2\epsilon Qra\sin^2\theta}{J}\frac{\partial S_a}{\partial_\phi}+\frac{e^2Q^2r^2a^2\sin^4\theta}{\rho^2J^2}\right]\right\}+4\lambda\rho^4\Upsilon_0\\&+\left(1+2a'+2\lambda c_r^2\right)A_r\rho^2\left(\frac{\partial S_a}{\partial_r}\right)^2+\left(1+2a'+2\lambda c_\theta^2\right)\rho^2\left(\frac{\partial S_a}{\partial_\theta}\right)^2=0\end{aligned} \tag{30}$$

where

$$\begin{aligned}\Upsilon_0=&u^t\left(\frac{\partial S_a}{\partial t}-eA_t\right)\left(u^r\frac{\partial S_a}{\partial r}+u^\theta\frac{\partial S_a}{\partial\theta}\right)+u^r\frac{\partial S_a}{\partial r}\left[u^\theta\frac{\partial S_a}{\partial\theta}+u^\phi\left(\frac{\partial S_a}{\partial\phi}-eA_\phi\right)\right]\\&+u^\theta u^\phi\left(\frac{\partial S_a}{\partial\theta}\right)\left(\frac{\partial S_a}{\partial\phi}-eA_\phi\right)+2b'\gamma^5-m^2\end{aligned} \tag{31}$$

In Equation (30), the sum of the four terms related to $\frac{1}{\Delta_\theta\sin^2\theta\rho^2}$ is zero. Multiplying both sides of this equation by $\frac{1}{\rho^2}$ yields

$$
\begin{aligned}
&\left(1+2a+2\lambda c_t^2\right)J^2\left\{\left[\frac{\left(a^2-(n-a\cos\theta)^2\right)^2}{a^2\Delta_\theta\sin^2\theta}-\frac{\left(r^2+a^2\right)^2}{\Delta_r}\right]\left[\left(\frac{\partial S_a}{\partial t}\right)^2+\frac{2eQr}{\rho^2 J}\frac{\partial S_a}{\partial t}\right]\right.\\
&+2\left[\frac{a\left(r^2+a^2\right)}{\Delta_r}+\frac{a^2-(n-a\cos\theta)^2}{\Delta_\theta a\sin^2\theta}\right]\left[\left(\frac{\partial S_a}{\partial t}\right)\left(\frac{\partial S_a}{\partial \phi}\right)-\frac{eQra\sin^2\theta}{\rho^2 J}\frac{\partial S_a}{\partial t}+\frac{eQr}{\rho^2 J}\frac{\partial S_a}{\partial \phi}\right]\\
&\left.+\left(\frac{1}{\Delta_\theta\sin^2\theta}-\frac{a^2}{\Delta_r}\right)\left[\left(\frac{\partial S_a}{\partial t}\right)^2-\frac{2eQra\sin^2\theta}{\rho^2 J}\frac{\partial S_a}{\partial \phi}\right]-\frac{e^2Q^2r^2\left(r^2+a^2+a^2\sin^2\theta\right)^2}{\Delta_r\rho^4 J^2}\right\}\\
&+\left(1+2a+2\lambda c_r^2\right)A_r\left(\frac{\partial S_a}{\partial r}\right)^2+\left(1+2a+2\lambda c_\theta^2\right)\left(\frac{\partial S_a}{\partial \theta}\right)^2+4\lambda\rho^2 Y_0=0
\end{aligned}
\tag{32}
$$

Separate the variables in the equation, letting

$$
S_a=-wt+R(r)+\Theta(\theta)+J_0\phi \tag{33}
$$

where w is the energy of the particle and J_0 is the component of the particle's generalized momentum in the ϕ direction, with J_0 being a constant. The ability to separate S_a into the form of Equation (33) is mainly due to the existence of two Killing vectors in the HNUTKN spacetime, namely $\left(\frac{\partial}{\partial t}\right)^\alpha$ and $\left(\frac{\partial}{\partial \phi}\right)^\alpha$. Substituting Equation (33) into Equation (32) and multiplying both sides of this equation by Δ_r, noting that $\Delta_r|_{r\to r_H}=0$, we obtain the form of the radial motion equation for the particle at the event horizon of this black hole as

$$
\begin{aligned}
\left[\Delta_r^2\left(\frac{dR}{dr}\right)^2\right]_{r\to r_H}&=\frac{J^2\left(1+2a'+2\lambda c_t^2\right)}{1+2a'+2\lambda c_r^2}\left[\left(r_H^2+a^2\right)^2\left(\frac{\partial S_a}{\partial t}\right)^2+2a\left(r^2+a^2\right)\frac{\partial S_a}{\partial t}\frac{\partial S_a}{\partial \phi}+a^2\left(\frac{\partial S_a}{\partial \phi}\right)^2\right.\\
&\left.+\frac{2eQr\left(r^2+a^2\right)^2}{J\left(r^2+a^2+n^2\right)}\frac{\partial S_a}{\partial t}+\frac{2aeQr\left(r^2+a^2\right)}{J\left(r^2+a^2+n^2\right)}\frac{\partial S_a}{\partial \phi}+\frac{e^2Q^2r^2\left(r^2+a^2\right)^2}{J^2\left(r^2+a^2+n^2\right)^2}\right]\\
&=\frac{J^2\left(1+2a'+2\lambda c_t^2\right)}{1+2a'+2\lambda c_r^2}\left[\left(r_H^2+a^2\right)\frac{\partial S_a}{\partial t}+a\frac{\partial S_a}{\partial \phi}+\frac{eQr\left(r_H{}^2+a^2\right)}{J\left(r_H{}^2+a^2+n^2\right)}\right]^2\\
&=\frac{J^2\left(1+2a'+2\lambda c_t^2\right)}{1+2a'+2\lambda c_r^2}\left(r_v{}^2+a^2\right)^2(\omega-\omega_0)^2
\end{aligned}
\tag{34}
$$

where

$$
\begin{aligned}
\omega_0&=\frac{aJ_0}{r_H^2+a^2}+\frac{eQr_H}{J\left(r_H^2+a^2+n^2\right)^2}\\
&=\frac{aJ_0J\left(r_H^2+a^2+n^2\right)+eQr_H\left(r_H^2+a^2\right)}{J\left(r_H^2+a^2\right)\left(r_H^2+a^2+n^2\right)}
\end{aligned}
\tag{35}
$$

Based on Equation (34), it can be inferred that the radial component $R(r)$ of the particle's generalized momentum satisfies the following equation at the event horizon of this black hole:

$$
\left.\frac{\mathrm{d}R^\pm}{\mathrm{d}r}\right|_{r\to r_H}=\pm\frac{\left(r_H^2+a^2\right)J}{\Delta_r|_{r\to r_H}}\left(\frac{1+2a'+2\lambda c_t^2}{1+2a'+2\lambda c_r^2}\right)^{\frac{1}{2}}(\omega-\omega_0). \tag{36}
$$

According to the residue theorem, from Equation (36), we obtain the radial component of the particle's generalized momentum as

$$
R^\pm|_{r\to r_H}=\pm\mathrm{i}\pi\frac{J\left(r_H^2+a^2\right)}{2\left[r_H-m+\frac{1}{3}\Lambda r_H\left(a^2-n^2\right)\right]}\left(\frac{1+2a'+2\lambda c_t^2}{1+2a'+2\lambda c_r^2}\right)^{\frac{1}{2}}(\omega-\omega_0) \tag{37}
$$

Let $R^\pm|_{r\to r_H}=R^\pm$; based on the WKB approximation theory and the black hole quantum tunneling radiation theory, the expression for the quantum tunneling rate of spin-1/2 fermions at the event horizon r_H of the black hole is

$$\begin{aligned}\Gamma \sim \exp\left(-2\,\mathrm{ImS}\,S_a^{\pm}\right) &= \exp\left(-2\,\mathrm{Im}\,R^{\pm}\right)\\ &= \exp\left[-\frac{2\pi J\left(r_H^2+a^2\right)}{r_H-m+\frac{1}{3}\Lambda r_H\left(a^2-n^2\right)}\left(\frac{1+2a'+2\lambda c_t^2}{1+2a'+2\lambda c_r^2}\right)^{\frac{1}{2}}(\omega-\omega_0)\right]\\ &= \exp\left(-\frac{\omega-\omega_0}{T_H}\right)\end{aligned} \tag{38}$$

where,

$$T_H = \frac{r_H-m+\frac{1}{3}\Lambda r_H\left(a^2-n^2\right)}{2\pi J\left(r_H^2+a^2\right)}\left(\frac{1+2a'+2\lambda c_r^2}{1+2a'+2\lambda c_t^2}\right)^{\frac{1}{2}} \tag{39}$$

Here, T_H is the Hawking temperature at the event horizon of the HNUTKN black hole, and the expression for this Hawking temperature contains Lorentz-breaking correction terms. The coefficient for the CFJ correction term is a', and the coefficient for the aether-like correction term is λ. The coefficient b' for the Chiral correction term does not appear in the expressions for Γ and T_{H}. This is due to $\Delta_r(r_H)\left(2b'\gamma_0^5-m^2\right)=0$ in Equation (32), so b' and m do not appear in the expressions for Γ and T_H. For fermions with any other spin, the coefficient b' for the Chiral correction term also does not appear in the expressions for Γ and T_H. Equations (37)–(39) are all results obtained in the semiclassical theory. We can use ΔS_{BH} to represent the Bekenstein–Hawking entropy of this black hole and represent the change in Bekenstein–Hawking entropy of this black hole as ΔS_{BH}, so the expression for the quantum tunneling rate of this black hole (38) can be rewritten as

$$\Gamma = \mathrm{e}^{(\Delta S_{BH})} \tag{40}$$

The quantity S_{BH} is closely related to the first law of thermodynamics for black holes. The mathematical relationship between S_{BH} and the black hole mass M, black hole angular velocity Ω, black hole charge Q, and black hole Hawking temperature T_H can be expressed as

$$\begin{aligned}\mathrm{d}S_{BH} &= \frac{\mathrm{d}M-\Omega\mathrm{d}J-Q\mathrm{d}V}{T_H}\\ &= \frac{\mathrm{d}M-\Omega\mathrm{d}J-Q\mathrm{d}V}{T_h}\left(\frac{1+2a'+2\lambda c_t^2}{1+2a'+2\lambda c_r^2}\right)^{\frac{1}{2}}\end{aligned} \tag{41}$$

where T_h is the Hawking temperature of this black hole without Lorentz-breaking correction terms. The expression for T_h, as given by Equation (39), is as follows:

$$T_h = \frac{r_H-m+\frac{1}{3}\Lambda r_H\left(a^2-n^2\right)}{2\pi J\left(r_H^2+a^2\right)} \tag{42}$$

Therefore, if we denote the Bekenstein–Hawking entropy as S_{bh}, the integral form of Equation (41) can be rewritten as

$$S_{BH} = \left(\frac{1+2a'+2\lambda c_t^2}{1+2a'+2\lambda c_r^2}\right)^{\frac{1}{2}}\int \mathrm{d}S_{bh} \tag{43}$$

According to Equation (15), it can be inferred that the two-dimensional line element on the event horizon of this black hole is given by

$$\begin{aligned}\mathrm{d}\sigma^2 &= \frac{\rho^2}{\Delta_\theta}\mathrm{d}\theta^2 + \left[\frac{\Delta_\theta\sin^2\theta}{\rho^2 J^2}\left(r_H^2+a^2\right)^2 - \frac{\Delta_r(r_H)}{J^2\rho^2}\left(a-\frac{(n-a\cos\theta)^2}{a}\right)^2\right]\mathrm{d}\phi^2\\ &= \frac{\rho^2}{\Delta_\theta}\mathrm{d}\theta^2 + \frac{\Delta_\theta\sin^2\theta}{\rho^2 J^2}\left(r_H^2+a^2\right)^2\mathrm{d}\phi^2\end{aligned} \tag{44}$$

Therefore, it can be seen that the determinant $g_{\theta\phi}$ corresponding to the two-dimensional covariant metric here is

$$g_{\theta\phi} = \frac{\sin^2\theta}{J^2}\left(r_H^2 + a^2\right)^2 \tag{45}$$

Thus, the area of the event horizon of this black hole is given by

$$A_{EH} = \int \sqrt{g_{\theta\phi}}\mathrm{d}\theta\mathrm{d}\phi = \int_0^{2\pi}\int_0^{\pi} \frac{\sin^2\theta}{J^2}\left(r_H^2 + a^2\right)^2 \mathrm{d}\theta\mathrm{d}\phi = \frac{4\pi}{J}\left(r_H^2 + a^2\right) \tag{46}$$

The Bekenstein–Hawking entropy of this black hole is proportional to the area of its event horizon and can be expressed as

$$S_{bh} = \frac{\pi}{J}\left(r_H^2 + a^2\right) \tag{47}$$

Therefore, Equation (43) can be rewritten as

$$S_{BH} = \frac{\pi}{J}\left(r_H^2 + a^2\right)\left(\frac{1 + 2a' + 2\lambda c_t^2}{1 + 2a' + 2\lambda c_r^2}\right)^{\frac{1}{2}} \tag{48}$$

This is the Bekenstein–Hawking entropy of the HNUTKN black hole in the framework of radius-metric theory, after Lorentz-breaking theory modifications. This is still a relevant result within the framework of radius-metric theory. Using quantum perturbation theory, we can make quantum corrections to the Bekenstein–Hawking entropy of this black hole [25–28]. In the framework of $\hbar$ perturbation theory, the energy E and radial action $R(r)$ of the quantum tunneling radiation particles can be expressed as

$$E = E_0 + \hbar E_1 + \hbar^2 E_2 + \cdots \tag{49}$$

$$\tilde{R}(r) = R_0(r) + \hbar R_1(r) + \hbar^2 R_2 + \cdots \tag{50}$$

$E_0 = (\omega - \omega_0)$ corresponds to the energy in Equations (37) and (38). $R_0(r)$ satisfies Equations (34) and (36), and the expression for $R_0^{\pm}$ is given in Equation (37), which can be considered as $R_0 \to R_0^{\pm}$, where $R_0^{\pm} = \pm\frac{i\pi J(r_H^2+a^2)}{2r_H - m + \frac{1}{3}\Lambda r_H(a^2-n^2)}\left(\frac{1+2a'+2\lambda c_t^2}{1+2a'+2\lambda c_r^2}\right)^{\frac{1}{2}}(\omega - \omega_0)$. Therefore, by expanding Equations (49) and (50) for $\hbar^0, \hbar^1, \hbar^2, \cdots$ and corresponding to different powers of $\hbar$, we can obtain

$$\begin{aligned}
\hbar^0 : \left(\frac{\mathrm{d}R_0}{\mathrm{d}r}\right)^2\Bigg|_{r\to r_H} &= \frac{J^2\left(r_H^2 + a^2\right)^2}{\Delta_r^2|_{r\to r_H}}\left(\frac{1 + 2a' + 2\lambda c_t^2}{1 + 2a' + 2\lambda c_r^2}\right)E_0^2, \\
\hbar^1 : \left(\frac{\mathrm{d}R_1}{\mathrm{d}r}\right)^2\Bigg|_{r\to r_H} &= \frac{J^2\left(r_H^2 + a^2\right)^2}{\Delta_r{}^2|_{r\to r_H}}\left(\frac{1 + 2a' + 2\lambda c_t^2}{1 + 2a' + 2\lambda c_r^2}\right)E_1{}^2.
\end{aligned} \tag{51}$$

From the equations satisfied by the radial actions corresponding to $\hbar^n$, it can be inferred that there exists a certain proportional relationship between $R_i^{\pm}$ and $R_0^{\pm}$. Therefore, the total energy E can be rewritten as

$$E = E_0 + \sum_{i=1}^{\infty}\frac{\alpha_i\hbar^i}{S_{bh}}E_0 \tag{52}$$

The total radial action can be rewritten as

$$\tilde{R}^{\pm} = \left(1 + \sum_{i=1}^{\infty}\frac{\beta_i\hbar^i}{S_{bh}}\right)R_0^{\pm} \tag{53}$$

The proportionality coefficients α_i and β_i in Equations (52) and (53) are both dimensionless. To modify the entropy of this black hole, we can express the scaling factors of $R_i^{\pm}$ and $R_0^{\pm}$ as $\frac{\beta_i \hbar^i}{S_{bh}}$. Therefore, according to the first law of black hole thermodynamics, the equation satisfied by the Bekenstein–Hawking entropy $\widetilde{S}_{BH}$ of this black hole with $\hbar$ perturbation corrections is approximately

$$\begin{aligned} d\widetilde{S}_{BH} &= \frac{dM - \Omega dJ - QdV}{T_H}\left(1 + \sum_{i=1}^{\infty}\frac{\beta_i \hbar^i}{S_{bh}}\right)^{-1} \\ &= \frac{dM - \Omega dJ - QdV}{T_h}\left(\frac{1+2a'+2\lambda c_t^2}{1+2a'+2\lambda c_r^2}\right)^{\frac{1}{2}}\left(1 - \sum_{i=1}^{\infty}\frac{\beta_i \hbar^i}{S_{bh}}\right) \end{aligned} \tag{54}$$

where T_h is the Hawking temperature at the event horizon of the black hole before Lorentz-breaking corrections are applied. Rewrite Equation (39) as

$$T_H = T_h\left(\frac{1+2a'+2\lambda c_r^2}{1+2a'+2\lambda c_t^2}\right) \tag{55}$$

Integrate both sides of Equation (54), noting $T_h d\,S_{\mathrm{bh}} = dM - \Omega dJ - QdV$, and we obtain

$$\begin{aligned} \widetilde{S}_{BH} &= \int dS_{bh}\left(\frac{1+2a'+2\lambda c_t^2}{1+2a'+2\lambda c_r^2}\right)^{\frac{1}{2}}\left(1 - \sum_{i=1}^{\infty}\frac{\beta_i \hbar^i}{S_{bh}}\right) \\ &= \left(\frac{1+2a'+2\lambda c_t^2}{1+2a'+2\lambda c_r^2}\right)^{\frac{1}{2}}\left(S_{bh} - \beta_i \hbar^i \ln S_{bh} + \cdots\right) \end{aligned} \tag{56}$$

The first term on the right-hand side of this equation represents the Bekenstein–Hawking entropy in the semiclassical theory of gravity, while the second term and subsequent terms represent the results of quantum perturbation theory corrections. It is evident that the black hole entropy after quantum corrections is not only dependent on the results of semiclassical theory but also on the Planck constant.

4. Discussion

Based on the research above, the Bekenstein–Hawking entropy of this black hole, denoted as S_{bh} in Equation (47), represents the entropy without any modifications. When considering the corrections due to Lorentz-breaking fermionic Einstein–aether theory, the Bekenstein–Hawking entropy of this black hole, denoted as S_{BH}, is derived as shown in Equation (48). Furthermore, incorporating quantum perturbation correction theories leads to the Bekenstein–Hawking entropy of this black hole, denoted as $\widetilde{S}_{BH}$, as given in Equation (56). The investigation of Bekenstein–Hawking entropy for different black holes remains a meaningful endeavor, emphasizing the ongoing relevance of research into topics such as black hole entropy, black hole information, and black hole thermodynamics. Equation (14) is a modified form of the dynamics equation for spin-1/2 fermions in semiclassical theory, which includes Lorentz-breaking correction terms. By studying this equation in the spacetime of the black hole, we obtain Hawking temperature T_H and Bekenstein-Hawking entropy S_{BH} of the HNUTKN black hole, which are results of semiclassical theory that include the effects of Lorentz-breaking corrections. Equations (53) and (56) both consider the results after quantum corrections, which go beyond semiclassical theory. Clearly, the entropy of this black hole is not only related to a', λ, C_t, and C_r, but also to $\hbar^i$. Equation (56) is the modified expression for the entropy of this black hole. The black hole studied in this paper possesses the characteristics of a general stationary black hole. Therefore, the series of results obtained above include relevant results for specific cases such as Kerr black holes and Kerr–Newman–de Sitter black holes. The HNUTKN black hole also has a cosmological horizon. If we want to study the quantum tunneling radiation characteristics at the cosmological horizon of the black hole, we only need to replace r_H with r_C in the research methods and the series of results obtained above. From the perspective of quantum tunneling radiation, the spatial region that the radiation reaches is $r > r_H$ at the event horizon of this black hole, while the spatial region that the radiation reaches is $r < r_C$ at the cosmological horizon of this black hole. Further discussion is needed on the physical significance of Equation (35). From the perspective of quantum tunneling radiation, ω_0 represents the chemical potential, which is formed by the rotation and charge of the black hole. It includes the electromagnetic potential and the rotational potential, thus determining the quantum tunneling radiation characteristics of this black hole. From the perspective of quantum non-thermal radiation, ω_0 is the maximum value of the particle Dirac energy level crossing, and this maximum value determines the condition for quantum non-thermal radiation to occur as $m < \omega \leqslant \omega_0$. If we want to study the quantum non-thermal radiation of this black hole, the mathematical expression of the Dirac energy level must include contributions from the Chiral correction term b'.

It should be further noted that the above research method and conclusions are specific to the HNUTKN black hole, and, for different types of black holes and fermions with different spins, specific analyses and research are required. For bosons, the above method cannot be used to study the modification of black hole entropy, and the modification of the scalar field action by Lorentz-breaking theory should be considered first. After modifying the boson dynamic equation on this basis, the particular significance of the black hole entropy can be analyzed.

Author Contributions: Writing—original draft, X.T.; Writing—review & editing, C.W. and S.-Z.Y. All authors have read and agreed to the published version of the manuscript.

Funding: This work is Supported by National Natural Science Foundation of China (Grant No. 12373109).

Institutional Review Board Statement: Not applicable.

Data Availability Statement: The data presented in this study are available on request from the corresponding author.

Conflicts of Interest: The authors declare no conflict of interest.

References

1. Gasperini, M. A thermal interpretation of the cosmological constant. *Class. Quantum Grav.* **1988**, *5*, 521. [CrossRef]
2. Ahmed, M. Hawking radiation of Dirac particles in the hot NUT-Kerr-Newman spacetime. *Phys. Lett. B* **1991**, *258*, 318. [CrossRef]
3. Parikh M.K.; Wilczek, F. Hawking Radiation As Tunneling. *Phys. Rev. Lett.* **2000**, *85*, 5042. [CrossRef]
4. Parikh, M.K. Energy Conservation and Hawking Radiation. *arXiv* **2004**, arXiv:hep-th/0402166.
5. Parikh, M.K. A Secret Tunnel through The Horizon. *Int. J. Mod. Phys. D* **2004**, *13*, 2351–2354. [CrossRef]
6. Feng, Z.W.; Li, H.L.; Zu, X.T.; Yang, S.Z. Corrections to the thermodynamics of Schwarzschild-Tangherlini black hole and the generalized uncertainty principle. *arXiv* **2015**, arXiv:1512.09219.
7. Hooft, G.T. On the quantum structure of a black hole. *Nucl. Phys. B* **1985**, *256*, 727. [CrossRef]
8. Li, X.; Zhao, Z. Entropy of an Extreme Reissner-Nordström Black Hole. *Gen. Rel. Grv.* **2002**, *34*, 255.
9. Li, X.; Zhao, Z. Entropy of a Vaidya black hole. *Phys. Rev. D* **2000**, *62*, 104001.
10. Liu, W.; Zhao, Z. Entropy of the Dirac field in a Kerr-Newman black hole. *Phys. Rev. D* **2000**, *61*, 063003.
11. Kruglov, S.I. Modified wave equation for spinless particles and its solutions in an external magnetic field. *Mod. Phys. Lett. A* **2013**, *28*, 1350014. [CrossRef]
12. Jacobson, T.; Liberati, S.; Mattingly, D. A strong astrophysical constraint on the violation of special relativity by quantum gravity. *Nature* **2003**, *424*, 1019. [CrossRef] [PubMed]
13. Amelino-Camelia, G. Phenomenology of Planck-scale Lorentz-symmetry test theories. *New J. Phys.* **2004**, *6*, 188. [CrossRef]
14. Magueijo, J.; Smolin, L. Lorentz invariance with an invariant energy scale. *Phys. Rev. Lett.* **2002**, *88*, 190403. [CrossRef]
15. Ellis, J.R.; Mavromatos, N.E.; Sakharov, A.S. Synchrotron radiation from the Crab Nebula discriminates between models of space–time foam. *Astropart. Phys.* **2004**, *20*, 669. [CrossRef]
16. Carleo, A.; Lambiase, G.; Mastrototaro, L. Energy extraction via magnetic reconnection in Lorentz breaking Kerr–Sen and Kiselev black holes. *Eur. Phys. J. C* **2022**, *82*, 776. [CrossRef]
17. Yang, S.Z.; Lin, K.; Li, J.; Jiang, Q.Q. Phenomenological Aspects of Quantum Gravity and Modified Theories of Gravity *Adv. High Energy Phys.* **2016**, *2016*, 7058764.

18. Tan, X.; Zhang, J.; Li, R. Correction of Lorentz breaking theory to quantum tunneling radiation and entropy of black hole in gravity's rainbow. *Phys. Scr.* **2023**, *98*, 105015. [CrossRef]
19. Tan, X.; Liu, Y.Z.; Liu, Z.E.; Sha, B.; Zhang, J.; Yang, S.Z. The solution of a modified Hamilton–Jacobi equation with Lorentz-violating scalar field. *Mod. Phys. Lett. A* **2020**, *35*, 2050168. [CrossRef]
20. Carroll, S.M.; Field, G.B.; Jackiw, R. Limits on a Lorentz and parity-violating modification of electrodynamics. *Phys. Rev. D* **1990**, *41*, 1231. [CrossRef]
21. Jackiw, R.; Kostelecký, V.A. Radiatively Induced Lorentz and CPT Violation in Electrodynamics. *Phys. Rev. Lett.* **1999**, *82*, 3572. [CrossRef]
22. Coleman, S.; Glashow, S.L. High-energy tests of Lorentz invariance. *Phys. Rev. D* **1999**, *59*, 116008. [CrossRef]
23. Gomes, M.; Nascimento, J.R.; Petrov, A.Y.; Da Silva, A.J. Aetherlike Lorentz-breaking actions. *Phys. Rev. D* **2010**, *81*, 045018. [CrossRef]
24. Nascimento, J.R.; Petrov, A.Y.; Marat Reyes, C. Lorentz-breaking theory with higher derivatives in spinor sector. *Phys. Rev. D* **2015**, *92*, 045030. [CrossRef]
25. Carter, B. The commutation property of a stationary. *Commun. Math. Phys.* **1970**, *17*, 233–238. [CrossRef]
26. Banerjee, R.; Majhi, B.R.; Samanta, S. Noncommutative black hole thermodynamics. *Phys. Rev. D* **2008**, *77*, 124035. [CrossRef]
27. Banerjee, R.; Majhi, B.R. Connecting anomaly and tunneling methods for hawking effect through Chirality. *Phys. Rev. D* **2009**, *79*, 394. [CrossRef]
28. Wang, C.; Tan, X.; Zhang, J.; Li, R.; Yang, S.Z. Correction of Kerr-Sen Black Hole Temperature and Entropy by Lorentz Invariance Violation. *Phys. Scr.* **2024**, *99*, 035009. [CrossRef]

entropy

MDPI

Article

The Hyperbolically Symmetric Black Hole

Luis Herrera [1,*,†] **and Louis Witten** [2,†]

[1] Instituto Universitario de Física Fundamental y Matemáticas, Universidad de Salamanca, 37007 Salamanca, Spain
[2] Independent Researcher, Ringhouse-180 E 1. Jefferson Road, Rockville, MD 20847, USA; lwittenw@gmail.com
* Correspondence: lherrera@usal.es
† These authors contributed equally to this work.

Abstract: We describe some properties of the hyperbolically symmetric black hole (hereafter referred to as the $HSBH$) proposed a few years ago. We start by explaining the main motivation behind such an idea, and we determine the main differences between this scenario and the classical black hole (hereafter referred to as the CBH) scenario. Particularly important are the facts that, in the $HSBH$ scenario, (i) test particles in the region inside the horizon experience a repulsive force that prevents them from reaching the center, (ii) test particles may cross the horizon outward only along the symmetry axis, and (iii) the spacetime within the horizon is static but not spherically symmetric. Next, we examine the differences between the two models of black holes in light of the Landauer principle and the Hawking results on the eventual evaporation of the black hole and the paradox resulting thereof. Finally, we explore what observational signature could be invoked to confirm or dismiss the model.

Keywords: black holes; exact solutions; general relativity

1. *HSBH* Versus *CBH*

Before contrasting the $HSBH$ with the CBH, we first present the general arguments leading us to propose the $HSBH$ model [1].

1.1. Why the HSBH Model?

As is well known in the CBH scenario, the spacetime outside the horizon is described by the Schwarzschild solution [2], whose line element in polar coordinates reads for $R > 2M$ (in relativistic units and with signature +2) as follows:

$$\begin{aligned} ds^2 &= -\left(1-\frac{2M}{R}\right)dt^2 + \frac{dR^2}{\left(1-\frac{2M}{R}\right)} + R^2 d\Omega^2, \\ d\Omega^2 &= d\theta^2 + \sin^2\theta d\phi^2, \end{aligned} \tag{1}$$

where M, which measures the total mass-energy of the source, is the only parameter of the solution.

Such a metric being static and spherically symmetric admits the four Killing vectors

$$\begin{aligned} \mathbf{X}_{(0)} = \partial_t, \quad \mathbf{X}_{(2)} = -\cos\phi\partial_\theta + \cot\theta\sin\phi\partial_\phi, \\ \mathbf{X}_{(1)} = \partial_\phi, \quad \mathbf{X}_{(3)} = \sin\phi\partial_\theta + \cot\theta\cos\phi\partial_\phi. \end{aligned} \tag{2}$$

Also, as is well known, the singularity appearing in (1) for $R = 2M$ is not a real physical singularity, and the solution may be analytically extended to the whole spacetime (including the region inside the horizon), as shown in [3–6].

However, the fact is that the spacetime within the horizon is necessarily non-static. Maintaining the static form of the Schwarzschild metric (in the whole spacetime) is incompatible with the possibility of removing the coordinate singularity appearing on the horizon in the line element [7]. This is related to the fact that static observers cannot be defined inside the horizon (see [8,9] for a discussion on this point). A simple way to see why this comes about consists of finding out the form of the null cone at any point inside the horizon by solving the equation

$$ds = 0. \tag{3}$$

The solution to the above equation shows that all the null rays generating the null cone (inside the horizon) converge to the center of symmetry, implying that anything within the null cone (including the massless particles along the null cone border) should reach the center (see [8–10] for a discussion on this point).

Although most of our colleagues are satisfied with this picture, we are not. Our concern is generated by the intuitive idea that any dynamical process should relax to an equilibrium state after some finite proper time. In such a case, we should be able to provide a global static description of the spacetime.

To the best of our knowledge, the first to call attention to this issue was Rosen [7], who proposed excluding the region $R < 2M$ because, in Rosen's words, "...the surface $R = 2M$ represents the boundary of physical space and should be regarded as an impenetrable barrier for particles and light rays" (see page 233 in [7]).

Sharing Rosen's concern about the physical properties of the region $R < 2M$, we proposed in [1] an alternative way to overcome the above-mentioned situation.

Thus, we proposed describing the region $R > 2M$ via the Schwarzschild solution (1); however, instead of excluding the region $R < 2M$, we assume that the spacetime in that region is defined by the line element (with signature -2)

$$\begin{aligned} ds^2 &= \left(\frac{2M}{R} - 1\right) dt^2 - \frac{dR^2}{\left(\frac{2M}{R} - 1\right)} - R^2 d\Omega^2, \\ d\Omega^2 &= d\theta^2 + \sinh^2\theta d\phi^2. \end{aligned} \tag{4}$$

The above metric is static; therefore, it admits the time-like Killing vector $\mathbf{X}_{(0)}$, and, furthermore, it admits the three Killing vectors

$$\begin{aligned} \mathbf{Y}_{(2)} &= -\cos\phi\partial_\theta + \coth\theta\sin\phi\partial_\phi, \\ \mathbf{Y}_{(1)} = \partial_\phi, \quad \mathbf{Y}_{(3)} &= \sin\phi\partial_\theta + \coth\theta\cos\phi\partial_\phi, \end{aligned} \tag{5}$$

implying that it is not spherically symmetric but hyperbolically symmetric.

The above three Killing vectors (5) define the hyperbolical symmetry. For applications of this kind of symmetry, see [11–33] and the references therein.

The situation may be summarized as follows: We have already seen that, when extending the Schwarzschild metric to the region inner to the horizon, which means keeping the spherical symmetry, one should abandon staticity; this is the picture for the CBH. However, we wish to keep staticity within the horizon. Accordingly, we propose abandoning sphericity, which leads to the $HSBH$ picture.

A short comment is in order at this point: In [1], we arrive at (4) from (1) through the transformation $\theta \to i\theta$. In doing so, the signature of the resulting line element (4) is different from the signature of (1). However, obviously, (4) is a static solution to the vacuum Einstein equations independently on its signature. In other words, the transformation mentioned above is just a technicality, and we may assume the region inside the horizon to be described by a hyperbolically symmetric metric with the same signature as the Schwarzschild line element describing the region outside the horizon. Thus, we do have a change in symmetry across the horizon but not necessarily a change in its signature.

1.2. Geodesics in HSBH

In [19], a general study of geodesics in the spacetime described by (4) is presented (see also [25]), leading to some interesting conclusions that highlight great differences between the behavior of a test particle in the $HSBH$ and the results emerging from the CBH, particularly the following:

- Inside the region $R < 2M$, the gravitational force appears to be repulsive.
- As a consequence of the above, test particles never reach the center.
- Unlike the CBH, test particles can cross the horizon outward, but only along the $\theta = 0$ axis.

Based on the last point above, it could be (rightly) argued that the object described by the $HSBH$ is not really "black" since matter may cross the horizon outwardly. However, in order to keep the link between the two scenarios (the $HSBH$ and CBH), we decide to keep the term "black".

The results above are supported by the expression of the four-acceleration (a^μ) of a static observer in the frame of (4). Indeed, a^μ represents the inertial radial acceleration that should be applied to the frame in order to keep the frame static by opposing the gravitational acceleration exerted on it.

Next, for a static observer, the four-velocity U^μ is proportional to the Killing time-like vector [8]. Then, for (4), we obtain

$$U^\mu = \left[\frac{1}{(\frac{2M}{r} - 1)^{1/2}}, 0, 0, 0\right], \tag{6}$$

producing for the four-acceleration $a^\mu \equiv U^\beta U^\mu_{;\beta}$ within the region inner to the horizon

$$a^\mu = \left[0, -\frac{M}{r^2}, 0, 0\right], \tag{7}$$

whereas for the region outside the horizon, defined by (1), the four-velocity vector of a static observer reads

$$U^\mu = \left[\frac{1}{(1 - \frac{2M}{r})^{1/2}}, 0, 0, 0\right], \tag{8}$$

producing for the four-acceleration

$$a^\mu = \left[0, \frac{M}{r^2}, 0, 0\right]. \tag{9}$$

From a simple inspection of (9), we see that the gravitational force in the region outer to the horizon is attractive, as it follows from the positive value of the acceleration (directed radially) in this region, while it is repulsive in the region inner to the horizon, as it follows from (7). The repulsive nature of gravitation in the spacetime described by (4) is characteristic of hyperbolical spacetimes and is at the origin of the peculiarities of the orbits within the horizon in the $HSBH$.

1.3. Flow of Information and Landauer Principle

The Landauer principle [34] asserts that, in the process of erasing one bit of information, some energy must be dissipated, a lower bound of which is given by

$$\triangle E = kT \ln 2, \tag{10}$$

where k is the Boltzmann constant, and T denotes the temperature of the environment (see also [35,36]).

Its relevance stems from the fact that it leads in a natural way to an "informational" reformulation of thermodynamics, which, in turn, allows for establishing a link between information theory and different branches of science [37–40].

The link between the Landauer principle and general relativity is discussed in detail in [41]. Here, we resort to some results found in that reference to carry on our discussion on the differences between the $HSBH$ and CBH.

Thus, a mass given by

$$M_{bit} = \frac{kT}{c^2} \ln 2, \tag{11}$$

was assigned to any bit of information, where c is the speed of light [42].

This issue (the mass of information) has been discussed by several authors (see [43–45] and the references therein).

Our goal in this subsection consists of contrasting the behavior of the information flow across the horizon in the gravitational field of a static mass in the $HSBH$ and CBH scenarios. To this end, first, we need an expression for the Landauer principle in the presence of a gravitational field.

The presence of a gravitational field in the weak field approximation produces a change in the total amount of minimal dissipated energy, which is given by [46]

$$\triangle E = \frac{kT\left(1 + \frac{\phi}{c^2}\right)}{c^2} \ln 2, \tag{12}$$

where ϕ denotes the Newtonian (negative) gravitational potential. Thus, in the presence of a gravitational field, the Landauer principle is modified by replacing the temperature T by the Tolman temperature [47], which, in the weak field limit, reads $T\left(1 + \frac{\phi}{c^2}\right)$.

Let us recall that, in presence of a gravitational field, thermodynamic equilibrium is ensured by the condition that the Tolman temperature is constant. This result, which is a consequence of the inertia of thermal energy, emerges naturally in the relativistic transport equations proposed in [48–50], and it should hold in any physically admissible, relativistic transport equation.

The generalization of (10) for a gravitational field of an arbitrary intensity reads

$$\Delta E = kT\sqrt{|g_{tt}|} \ln 2, \tag{13}$$

whereas the expression for the mass of a bit becomes

$$M_{bit} = \frac{kT\sqrt{|g_{tt}|}}{c^2} \ln 2, \tag{14}$$

where g_{tt} is the tt component of the metric tensor, and the fact is used that the Tolman temperature is defined by $T\sqrt{|g_{tt}|}$, which, in the weak field limit, becomes $T(1 + \frac{\phi}{c^2})$.

Let us now consider the CBH scenario.

In this case, if we admit that a bit of information is endowed with a mass according to (14), then, since nothing can cross the horizon outwardly, it follows that no information can cross the horizon outwardly. This conclusion agrees with the results obtained by Hawking about the radiation of a CBH due to quantum effects [51] and its eventual evaporation, as well as with the fact that such radiation is completely thermal (i.e., it conveys no information) [52]), which leads to a contradiction known as the information loss paradox.

Indeed, the Hawking result implies that a pure quantum state evolves into a mixed state (the thermal radiation), which, of course, contradicts the principle of unitary evolution. Since the Hawking radiation is thermal, it appears that all information about the collapsing object is lost forever.

In spite of the intense research work devoted to this issue in recent decades, no satisfactory resolution of this quandary has yet been proposed.

Let us next consider the $HSBH$ scenario.

As mentioned before, in such a case, the crossing of massive particles through the horizon outwardly is allowed along the $\theta = 0$ axis, thereby implying the existence of a flux of information from the inside of the horizon to the outside. Thus, in this scenario, no information loss paradox appears, since even if the Hawking radiation is thermal, information may leave the collapsed object.

Furthermore, in this latter picture, the energy flow along the axis $\theta = 0$ associated with the change in information within the horizon could be, in principle, observable.

Thus, invoking the Landauer principle in the study of the global picture of the Schwarzschild black hole may lead to observable consequences, which could help to elucidate the quandary about the real nature of the Schwarzschild black hole.

2. Observational Evidences

Finally, we would like to mention some observational projects that could be determinant to dismiss or confirm the $HSBH$ scenario.

The most important of these endeavors seems to be a study based on the information provided by the Event Horizon Telescope (EHT) Collaboration [53–55], which provides observations of shadow images of the gravitationally collapsed objects at the center of the elliptical galaxy $M87$ and at the center of the Milky Way Galaxy. Such observations are expected to provide important data on strong fields [56–58], which could be used to obtain constraints for the parameters of the solutions describing the geometry surrounding compact objects, e.g., black hole spacetimes in modified and alternative theories of gravity [59–63], naked singularities, and classical GR black holes with hair or immersed in matter fields [64–67].

It should be noted that, although the region exterior to the horizon is described by the usual Schwarzschild metric, we might expect some imprints on the shadows from the material ejected along the symmetry axis in the $HSBH$.

Another possible source of information to support the $HSBH$ picture could be the observation and modeling of extragalactic relativistic jets.

These are highly energetic phenomena that have been observed in many systems (see [68–71] and the references therein), usually associated with the presence of a compact object. One of their characteristic properties is a high degree of collimation (besides the extremely high energies involved).

So far, no consensus has been reached concerning the basic mechanism explaining these two features of jets (collimation and high energies); therefore, it is legitimate to speculate that the $HSBH$ could be considered a possible engine behind the jets.

Indeed, let us recall that, in the $HSBH$ scenario, test particles may cross the horizon outward, but only along the $\theta = 0$ axis, thereby explaining the collimation. However, as it follows from (7), the strength of the repulsive gravitational force acting on the particle as

$r \to 0$ increases as $\frac{1}{r^2}$, thereby explaining the high energies of the particles bouncing back from regions close to $r = 0$.

Obviously, at this level of generality, this is just speculation. A solid theory for relativistic jets would require a much more detailed setup based on astronomical observations.

3. Discussion and Conclusions

Above, we exposed the main features of the $HSBH$ and contrasted them with those of the CBH. We started by explaining that our main motivation for proposing the $HSBH$ was the need to describe the region interior to the horizon by a static spacetime, which, in turn, forced us to abandon the assumption of spherical symmetry in that region. The analysis of the geodesic structure within the horizon exhibited profound differences from the corresponding structure in the CBH.

As we showed, such differences are related to the fact that the nature of gravity within the horizon happens to be repulsive, as a consequence of which test particles never reach the center. This feature is perhaps the most relevant difference between the $HSBH$ and CBH.

The origin of the repulsive nature of gravitation within the horizon should be found in the presence of negative mass (energy). This is indeed the case, as shown in the studies of fluid distributions endowed with hyperbolical symmetry (see [18,21,22,30] for details).

It is worth mentioning that the existence of negative mass (energy) in gravitational physics has a long and venerable history, at both the classical [10,72–77] and quantum levels [78–84].

We next invoked the Landauer principle to illustrate the differences between the $HSBH$ and CBH concerning the possible flow of information and the discussion of the Hawking results in both scenarios. It follows in a natural way that no information loss paradox appears in the $HSBH$.

Finally, we call attention to a new line of investigations involving observations of shadow images of the gravitationally collapsed outcome, aiming to identify the nature of different compact objects. This could provide arguments to support (or dismiss) the $HSBH$ picture. Also, the idea has been put forward about the possible use of the $HSBH$ as the engine of extragalactic jets.

Before concluding, we would like to highlight three important issues that we believe deserve further investigation:

- The two manifolds describing the inner and outer parts of the horizon do not match smoothly in the Darmois sense [85]. This implies that there is a shell on the horizon, whose physical and mathematical properties are described by the Israel conditions [86]. It would be interesting to delve deeper into this issue and find out how these properties affect the $HSBH$ scenario.
- It would be interesting to find out whether a "thermodynamic" approach *a la Bekenstein* may also be applied to the $HSBH$.
- The $HSBH$ scenario implies a change in symmetry across the horizon; is there any physical explanation for this (e.g., a phase transition)?

Author Contributions: Conceptualization, L.H., L.W.; methodology, L.H., L.W.; formal analysis, L.H., L.W.; writing—original draft preparation, L.H. writing—review and editing, L.H., L.W.; funding acquisition, L.H. All authors have read and agreed to the published version of the manuscript.

Funding: This work was partially supported by the Spanish Ministerio de Ciencia, Innovacion, under Research Project No. PID2021-122938NB-I00.

Institutional Review Board Statement: Not applicable.

Informed Consent Statement: Not applicable.

Data Availability Statement: Data are contained within the article.

Conflicts of Interest: The authors declare no conflicts of interest. The funders had no role in the design of the study; in the collection, analyses, or interpretation of data; in the writing of the manuscript; or in the decision to publish the results.

References

1. Herrera, L.; Witten, L. An alternative approach to the static spherically symmetric vacuum global solutions to the Einstein's equations. *Adv. High Energy Phys.* **2018**, *2018*, 3839103. [CrossRef]
2. Schwarzschild, K. Uber das Gravitationsfeld eines Massenpunktes nach der Einsteinschen Theorie. *Sitzungsberichte Königlich Preußischen Akad. Wiss.* **1916**, 189–196.
3. Lemaître, G. L'Univers en expansion. *Ann. Soc. Sci. Brux.* **1933**, *A 53*, 51.
4. Finkelstein, D. Past-Future Asymmetry of the Gravitational Field of a Point Particle. *Phys. Rev.* **1958**, *110*, 965. [CrossRef]
5. Kruskal, M.D. Maximal Extension of Schwarzschild Metric. *Phys. Rev.* **1960**, *119*, 1743. [CrossRef]
6. Israel, W. New Interpretation of the Extended Schwarzschild Manifold. *Phys. Rev.* **1966**, *143*, 1016. [CrossRef]
7. Rosen, N. The nature of the Schwarzschild singularity. In *Relativity*; Carmeli, M., Fickler, S.I., Witten, L., Eds.; Plenum Press: New York, NY, USA, 1970; pp. 229–258.
8. Caroll, S. *Spacetime and Geometry. An Introduction to General Relativity*; Addison Wesley: San Francisco, CA, USA, 2004; p. 246.
9. Rindler, W. *Relativity. Special, General and Cosmological*; Oxford University Press: New York, NY, USA, 2001; pp. 260–261.
10. Papapetrou, A. *Lectures on General Relativity*; D. Reidel, Dordrecht-Holland: Boston, MA, USA, 1974.
11. Harrison, B.K. Exact Three-Variable Solutions of the Field Equations of General Relativity. *Phys. Rev.* **1959**, *116*, 1285. [CrossRef]
12. Ellis, G. Dynamics of Pressure-Free Matter in General Relativity. *J. Math. Phys.* **1967**, *8*, 1171. [CrossRef]
13. Stephani, H.; Kramer, D.; MacCallum, M.; Honselaers, C.; Herlt, E. *Exact Solutions to Einsteins Field Equations*, 2nd ed.; Cambridge University Press: Cambridge, UK, 2003.
14. Gaudin, M.; Gorini, V.; Kamenshchik, A.; Moschella, U.; Pasquier, V. Gravity of a static massless scalar field and a limiting Schwarzschild-like geometry. *Int. J. Mod. Phys. D* **2006**, *15*, 1387–1399. [CrossRef]
15. Rizzi, L.; Cacciatori, S.L.; Gorini, V.; Kamenshchik, A.; Piattella, O.F. Dark matter effects in vacuum spacetime. *Phys. Rev. D* **2010**, *82*, 027301. [CrossRef]
16. Kamenshchik, A.Y.; Pozdeeva, E.O.; Starobinsky, A.A.; Tronconi, A.; Vardanyan, T.; Venturi, G.; Yu, S. Verno. Duality between static spherically or hyperbolically symmetric solutions and cosmological solutions in scalar-tensor gravity. *Phys. Rev. D* **2018**, *98*, 124028. [CrossRef]
17. Madler, T. On the affine-null metric formulation of General Relativity. *Phys. Rev. D* **2019**, *99*, 104048. [CrossRef]
18. Maciel, A.; Delliou, M.L.; Mimoso, J.P. New perspectives on the TOV equilibrium from a dual null approach. *Class. Quantum Gravity* **2020**, *37*, 125005. [CrossRef]
19. Herrera, L.; Di Prisco, A.; Ospino, J.; Witten, L. Geodesics of the hyperbolically symmetric black hole. *Phys. Rev. D* **2020**, *101*, 064071. [CrossRef]
20. Bhatti, M.Z.; Yousaf, Z.; Tariq, Z. Hyperbolically symmetric sources in Palatini $f(R)$ gravity. *Eur. Phys. J. C* **2021**, *81*, 1070. [CrossRef]
21. Herrera, L.; Di Prisco, A.; Ospino, J. Dynamics of hyperbolically symmetric fluids. *Symmetry* **2021**, *13*, 1568. [CrossRef]
22. Herrera, L.; Di Prisco, A.; Ospino, J. Hyperbolically symmetric static fluids: A general study. *Phys. Rev. D* **2021**, *103*, 024037. [CrossRef]
23. Herrera, L.; Di Prisco, A.; Ospino, J. Hyperbolically symmetric versions of Lemaitre-Tolman-Bondi spacetimes. *Entropy* **2021**, *23*, 1219. [CrossRef] [PubMed]
24. Bhatti, M.Z.; Yousaf, Z.; Hanif, S. Hyperbolically symmetric sources, a comprehensive study in $f(T)$ gravity. *Eur. Phys. J. Plus* **2022**, *137*, 65. [CrossRef]
25. Lim, Y. Motion of charged particles in spacetimes with magnetic fields of spherical and hyperbolic symmetry. *Phys. Rev. D* **2022**, *106*, 064023. [CrossRef]
26. Yousaf, Z.; Bhatti, M.; Khlopov, M.; Asad, H. A Comprehensive Analysis of Hyperbolical Fluids in Modified Gravity. *Entropy* **2022**, *24*, 150. [CrossRef]
27. Asad, H.; Yousaf, Z. Study of Anisotropic Fluid Distributed Hyperbolically in $f(R,T,Q)$ Gravity. *Universe* **2022**, *8*, 630. [CrossRef]
28. Yousaf, Z.; Nashed, G.; Bhatti, M.; Asad, H. Significance of Charge on the Dynamics of Hyperbolically Distributed Fluids. *Universe* **2022**, *8*, 337. [CrossRef]
29. Yousaf, Z.; Bhatti, M.; Khan, S. Analysis of hyperbolically symmetric fluid configurations in modified Gauss-Bonnet gravity. *Eur. Phys. J. C* **2022**, *82*, 1077. [CrossRef]
30. Herrera, L. Non static hyperbolically symmetric fluids. *Int. J. Mod. Phys. D* **2022**, *31*, 2240001. [CrossRef]

31. Herrera, L.; Di Prisco, A.; Ospino, J.; Carot, J. Quasi-hyperbolically symmetric γ-metric. *Entropy* **2023**, *25*, 1338. [CrossRef] [PubMed]
32. Yousaf, Z.; Bhatti, M.; Asad, H. Analytical models of hyperbolical gravitational sources. *Int. J. Mod. Phys. D* **2023**, *32*, 2350089. [CrossRef]
33. Carrasco, M.; Contreras, E.; Fuenmayor, E. Hyperbolic polytrope. *Ann. Phys.* **2025**, *473*, 169909. [CrossRef]
34. Landauer, R. Dissipation and heat generation in the computing process. *IBM J. Res. Dev.* **1961**, *5*, 183. [CrossRef]
35. Kish, L.B.; Granqvist, C.G. Electrical Maxwell demon and Szilard engine utilizing Johnson noise, measurement, logic and control. *PLoS ONE* **2012**, *7*, e46800. [CrossRef]
36. Brillouin, L. The negentropic principle of information. *J. Appl. Phys.* **1953**, *24*, 1152–1163. [CrossRef]
37. Bormashenko, E. The Landauer Principle: Re-Formulation of the Second Thermodynamics Law or a Step to Great Unification. *Entropy* **2019**, *21*, 918. [CrossRef]
38. Plenio, M.B.; Vitelli, V. The physics of forgetting: Landauer's erasure principle and information theory. *Contemp. Phys.* **2001**, *42*, 25–60. [CrossRef]
39. Bais, F.A.; Farmer, J.D. The physics of information. *arXiv* **2007**, arXiv:0708.2837v2. [CrossRef]
40. Bormashenko, E. Generalization of the Landauer Principle for Computing Devices Based on Many-Valued Logic. *Entropy* **2019**, *21*, 1150. [CrossRef]
41. Herrera, L. Landauer principle and general relativity. *Entropy* **2020**, *22*, 340. [CrossRef]
42. Herrera, L. The mass of a bit of information and the Brillouin's principle. *Fluc. Noise Lett.* **2014**, *13*, 1450002. [CrossRef]
43. Kish, L.B. Gravitational mass of information? *Fluct. Noise Lett.* **2007**, *7*, C51–C68. [CrossRef]
44. Kish, L.B.; Granqvist, C.G. Does information have mass. *Proc. IEEE* **2013**, *9*, 1895–1899. [CrossRef]
45. Vopson, M. The mass-energy-information equivalence principle. *AIP Adv.* **2019**, *9*, 095206. [CrossRef]
46. Daffertshoffer, A.; Plastino, A.R. Forgetting and gravitation: From Landauer's principle to Tolman temperature. *Phys. Lett. A* **2007**, *362*, 243–245. [CrossRef]
47. Tolman, R. On the weight of heat and thermal equilibrium in general relativity. *Phys. Rev.* **1930**, *35*, 904–924. [CrossRef]
48. Eckart, C. The thermodynamics of irreversible processes III. Relativistic theory of the simple fluid. *Phys. Rev.* **1940**, *58*, 919–924. [CrossRef]
49. Landau, L.D.; Lifshitz, E.M. *Fluid Mechanics*; Pergamon Press: London, UK, 1959.
50. Israel, W.; Stewart, J. Transient relativistic thermodynamics and kinetic theory. *Ann. Phys.* **1979**, *118*, 341–372. [CrossRef]
51. Hawking, S.W. Particles creation by black holes. *Commun. Math. Phys.* **1975**, *43*, 199–220. [CrossRef]
52. Hawking, S.W. Breakdown of predictability in gravitational collapse. *Phys. Rev. D* **1976**, *14*, 2460–2473. [CrossRef]
53. Sasada, M.; Event Horizon Telescope Collaboration. First M87 Event Horizon Telescope Results. I. The Shadow of the Supermassive Black Hole. *Astrophys. J. Lett.* **2019**, *875*, L1.
54. Akiyama, K.; Alberdi, A.; Alef, W.; Algaba, J.C.; Anantua, R.; Asada, K.; Azulay, R.; Bach, U.; Baczko, A.K.; Ball, D.; et al. First Sagittarius A Event Horizon Telescope Results. I. The Shadow of the Supermassive Black Hole in the Center of the Milky Way. *Astrophys. J. Lett.* **2022**, *930*, L12.
55. Vagnozzi, S.; Roy, R.; Tsai, Y.D.; Visinelli, L.; Afrin, M.; Allahyari, A.; Bambhaniya, P.; Dey, D.; Ghosh, S.G.; Joshi, P.S.; et al. Horizon-scale tests of gravity theories and fundamental physics from the Event Horizon Telescope image of Sagittarius *A*. *arXiv* **2022**, arXiv:2205.07787. [CrossRef]
56. Psaltis, D. Testing general relativity with the Event Horizon Telescope. *Gen. Relativ. Gravit.* **2019**, *51*, 137. [CrossRef]
57. Gralla, S.E. Can the EHT M87 results be used to test general relativity? *Phys. Rev. D* **2021**, *103*, 024023. [CrossRef]
58. Glampedakis, K.; Pappas, G. Can supermassive black hole shadows test the Kerr metric? *Phys. Rev. D* **2021**, *104*, L081503. [CrossRef]
59. Ghasemi-Nodehi, M.; Azreg-Ainou, M.; Jusufi, K.; Jamil, M. Shadow, quasinormal modes, and quasiperiodic oscillations of rotating Kaluza-Klein black holes. *Phys. Rev. D* **2020**, *102*, 104032. [CrossRef]
60. Jusufi, K.; Azreg-Ainou, M.; Jamil, M.; Wei, S.W.; Wu, Q.; Wang, A. Quasinormal modes, quasiperiodic oscillations, and the shadow of rotating regular black holes in nonminimally coupled Einstein-Yang-Mills theory. *Phys. Rev. D* **2021**, *103*, 024013. [CrossRef]
61. Liu, C.; Zhu, T.; Wu, Q.; Jusufi, K.; Jamil, M.; Azreg-Ainou, M.; Wang, A. Shadow and quasinormal modes of a rotating loop quantum black hole. *Phys. Rev. D* **2020**, *101*, 084001. [CrossRef]
62. Jusufi, K.; Jamil, M.; Chakrabarty, H.; Wu, Q.; Bambi, C.; Wang, A. Rotating regular black holes in conformal massive gravity. *Phys. Rev. D* **2020**, *101*, 044035. [CrossRef]
63. Afrin, M.; Kumar, R.; Ghosh, S.G. Parameter estimation of hairy Kerr black holes from its shadow and constraints from M87. *Mon. Not. R. Astron. Soc.* **2021**, *504*, 5927. [CrossRef]
64. Cardoso, V.; Gualtieri, L. Testing the black hole "no-hair" hypothesis. *Class. Quantum Gravity* **2016**, *33*, 174001. [CrossRef]

65. Kurmanov, E.; Boshkayev, K.; Giambo, R.; Konysbayev, T.; Luongo, O.; Malafarina, D.; Quevedo, H. Accretion Disk Luminosity for Black Holes Surrounded by Dark Matter with Anisotropic Pressure. *Astrophys. J.* **2022**, *925*, 210. [CrossRef]
66. Boshkayev, K.; Konysbayev, T.; Kurmanov, E.; Luongo, O.; Malafarina, D.; Mutalipova, K.; Zhumakhanova, G. Effects of non-vanishing dark matter pressure in the Milky Way Galaxy. *Mon. Not. R. Astron. Soc.* **2021**, *508*, 1543. [CrossRef]
67. Boshkayev, K.; Idrissov, A.; Luongo, O.; Malafarina, D. Accretion disc luminosity for black holes surrounded by dark matter. *Mon. Not. R. Astron. Soc.* **2020**, *496*, 1115. [CrossRef]
68. Blandford, R.D.; Rees, M.J. A "Twin-Exhaust" Model for Double Radio Sources. *Mon. Not. R. Astron. Soc.* **1974**, *169*, 395. [CrossRef]
69. Margon, B.A. Observations of SS 433. *Annu. Rev. Astron. Astrophys.* **1984**, *22*, 507. [CrossRef]
70. Sams, B.J.; Eckart, A.; Sunyaev, R. Near-infrared jets in the Galactic microquasar $GRS1915+105$. *Nature* **1996**, *382*, 47–49. [CrossRef]
71. Blandford, R.D. AGN Jets . *Astron. Soc. Pacif. Conf. Ser.* **2003**, *290*, 267.
72. Bondi, H. Negative Mass in General Relativity. *Rev. Mod. Phys.* **1957**, *29*, 423–428. [CrossRef]
73. Cooperstock, F.I.; Rosen, N. A nonlinear gauge-invariant field theory of leptons. *Int. J. Theor. Phys.* **1989**, *28*, 423–440. [CrossRef]
74. Bonnor, W.B.; Cooperstock, F.I. Does the electron contain negative mass? *Phys. Lett. A* **1989**, *139*, 442–444. [CrossRef]
75. Herrera, L.; Varela, V. Negative energy density and classical electron models. *Phys. Lett. A* **1994**, *189*, 11–14. [CrossRef]
76. Najera, S.; Gamboa, A.; Aguilar-Nieto, A.; Escamilla-Rivera, C. On Negative Mass Cosmology in General Relativity. *Astron. Astrophys.* **2021**, *651*, L13. [CrossRef]
77. Bormashenko, E. The effect of negative mass in gravitating systems. *Pramana* **2023**, *97*, 199. [CrossRef]
78. Farnes, J.S. A unifying theory of dark energy and dark matter: Negative masses and matter creation within a modified λ cdm framework. *Astron. Astrophys.* **2018**, *620*, A92. [CrossRef]
79. Capozziello, S.; Lobo, F.S.N.; Mimoso, J.P. Energy conditions in modified gravity. *Phys. Lett. B* **2014**, *730*, 280–283. [CrossRef]
80. Capozziello, S.; Lobo, F.S.N.; Mimoso, J.P. Generalized energy conditions in extended theories of gravity. *Phys. Rev. D* **2015**, *91*, 124019. [CrossRef]
81. Barcelo, C.; Visser, M. Twilight for the Energy Conditions? *Int. J. Mod. Phys. D* **2002**, *11*, 1553–1560. [CrossRef]
82. Kontou, E.A.; Sanders, K. Energy conditions in general relativity and quantum field theory. *Class. Quantum Gravity* **2020**, *37*, 193001. [CrossRef]
83. Pavsic, M. On negative energies, strings, branes, and braneworlds: A review of novel approaches. *Int. J. Mod. Phys. A* **2020**, *35*, 2030020. [CrossRef]
84. Hao, C.H.; Huang, L.X.; Su, X.; Wang, Y.Q. Emergence of negative mass in general relativity. *Eur. Phys. J. C* **2024**, *84*, 878. [CrossRef]
85. Darmois, G. *Memorial des Sciences Mathematiques*; Gauthier-Villars: Paris, France, 1927; p. 25.
86. Israel, W. Singular hypersurfaces and thin shells in general relativity. *Il Nuovo Cimento B* **1966**, *10*, 1. [CrossRef]

entropy

Article

Entropy Modifications of Charged Accelerating Anti-de Sitter Black Hole

Cong Wang [1,*], Jie Zhang [2] and Shu-Zheng Yang [3]

[1] College of Physics and Electronic Engineering, Qilu Normal University, Jinan 250300, China
[2] School of Arts and Sciences, Shanghai Dianji University, Shanghai 200240, China; zhangjie@126.com
[3] College of Physics and Astronomy, China West Normal University, Nanchong 637002, China; szyangcwnu@126.com
* Correspondence: cctianlan@163.com

Abstract: The Lorentz-breaking theory not only modifies the geometric structure of curved spacetime but also significantly alters the quantum dynamics of bosonic and fermionic fields in black hole spacetime, leading to observable physical effects on Hawking temperature and Bekenstein–Hawking entropy. This study establishes the first systematic theoretical framework for entropy modifications of charged accelerating Anti-de Sitter black holes, incorporating gauge-invariant corrections derived from Lorentz-violating quantum field equations in curved spacetime. The obtained analytical expression coherently integrates semi-classical approximations with higher-order quantum perturbative contributions. Furthermore, the methodologies employed and the resultant conclusions are subjected to rigorous analysis, establishing their physical significance for advancing fundamental investigations into black hole entropy.

Keywords: quintessence; Anti-de Sitter black hole; Lorentz-breaking; quantum tunneling radiation; Bekenstein–Hawking entropy

1. Introduction

This study aims to investigate the thermodynamic properties of charged accelerating Anti-de Sitter (AdS) black holes and to elucidate the physical implications of entropy corrections within the Lorentz-breaking theoretical framework. To deepen our understanding of black hole entropy and its physical implications, it is essential to first clarify the methodology and definition of black hole entropy research. When treating a black hole as a thermodynamic system, its entropy is closely related to the Hawking temperature at the event horizon, as dictated by the first law of black hole thermodynamics. From a quantum theoretical perspective, the entropy is found to be proportional to the area of the event horizon, measured in Planck units ($\propto l_p^2$). This suggests that black hole thermodynamics encompasses quantum gravitational effects that warrant further exploration. Additionally, the black hole area theorem states that the surface area of a black hole never decreases in the forward time direction, which is a direct manifestation of the second law of black hole thermodynamics and serves as an important physical demonstration of the "arrow of time" in gravitational physics. Since 't Hooft proposed the brick wall model in 1985, research into black hole entropy has significantly advanced [1]. This model posits that the entropy of a black hole is equivalent to the entropy of a quantum gas in thermal equilibrium outside the black hole. Subsequently, Zhao and colleagues improved 't Hooft's brick wall model by developing the membrane paradigm [2–5], which enables the calculation of entropy for non-stationary black holes and static or stationary black holes in non-thermal equilibrium states. The membrane paradigm reveals that black hole entropy fundamentally originates from the entropy of a two-dimensional membrane located at the black hole's surface (event horizon). Specifically, it represents the entropy of a two-dimensional surface formed by the intersection of a null hypersurface with spacelike hypersurfaces in three-dimensional space. This entropy arises from contributions of a quantum gas on the two-dimensional membrane and can be interpreted as the entropy of quantum states on the two-dimensional horizon. The membrane paradigm is applicable to studying the entropy of static, stationary, and non-stationary black holes. Research has demonstrated that black hole entropy is proportional to the area of the event horizon. General relativity and quantum field theory are both founded on Lorentz symmetry [6–8]. However, these theories cannot be unified at high energies due to the inherent non-renormalizability of Einstein's general relativistic theory of gravity. As a result, researchers have pursued multiple avenues to investigate topics in quantum gravity theory. To date, no satisfactory quantum gravity theory has emerged to address the quantization of gravity. Fundamental contradictions between gravitational and quantum theories persist, demanding further exploration. Questions related to the quantization of gravity remain a hot topic among astrophysicists, driving meaningful and in-depth studies into diverse gravitational theories. Despite its potential in explaining quantum gravitational effects, loop quantum gravity theory still faces challenges such as the lack of experimental verification and the complexity of mathematical computations. Another fundamental approach to constructing gravitational theories assumes that Lorentz symmetry is broken at high energies but restored at low energies, consistent with current experimental results. This insight has inspired researchers to incorporate Lorentz-breaking mechanisms into studies of gravitational renormalization, such as Horava–Lifshitz gravity and Einstein-aether gravity. These theories exhibit consistency in their low-energy limits. In recent years, the effects of Lorentz-breaking on the curved spacetime background of black holes and its influence on the dynamics of bosons and fermions in black hole spacetimes have become significant research topics. However, in the realm of black hole thermodynamics, the corrections to Hawking temperature and Bekenstein–Hawking entropy due to Lorentz-breaking require further investigation. Previous studies have explored the impact of Lorentz-breaking on the Bekenstein–Hawking entropy of uncharged accelerating Kinnersley black holes and its physical implications [9]. Nevertheless, research on entropy corrections for charged accelerating AdS black holes remains insufficient. Such studies could enhance our understanding of the thermodynamic evolution characteristics of arbitrary charged, rotating, accelerating, and non-stationary black holes. According to the first law of black hole thermodynamics, Bekenstein–Hawking entropy is closely related to thermodynamic parameters such as Hawking temperature. Therefore, investigating changes in black hole entropy necessitates exploring its quantum tunneling radiation characteristics and thermodynamic evolution properties. In this paper, based on the WKB semiclassical theory, black hole quantum tunneling radiation theory, and Lorentz-breaking theory, we study the thermodynamic properties of charged accelerating AdS black holes. We also analyze the effects of Lorentz-breaking and $\hbar$ perturbation theory on the quantum tunneling radiation rates of bosons and fermions, as well as the Bekenstein–Hawking entropy of these black holes.

In Section 2, we first investigate the modified form of the bosonic dynamical equations in the spacetime of a charged accelerating AdS black hole. Building upon this foundation, we analyze the corrected entropy of this charged accelerating AdS black hole and explore its physical implications. In Section 3, we extend our study to the modification of fermionic dynamical equations and examine the corresponding corrections to black hole entropy. Finally, in Section 4, we provide a detailed discussion of the methodologies employed in this work and the significance of the obtained results.

2. Research on Hawking Temperature and Bekenstein–Hawking Entropy of Charged Accelerating AdS Black Holes

In the study of string theory and quantum gravity theories, it has been found that the Lorentz dispersion relations require modifications at the Planck-scale level in high-energy domains [10–14]. Based on Einstein-aether gravity theory and Lorentz-breaking theory, the action S_b of the scalar field φ is modified as follows:

$$S_b = \frac{1}{2}\int d^4x\sqrt{-g}\left[\left(\tilde{\Box} + \frac{m^2}{\hbar^2}\right)\varphi^2 + \lambda\left(u^\mu\partial_\mu\varphi\right)^2\right] \tag{1}$$

According to the computational rules of index contraction, $(u^\mu\partial_\mu)^2$ should be expressed as $u^\mu\partial_\mu u^\nu\partial_\nu$. In Equation (1), the operator $\tilde{\Box}$ represents the d'Alembertian operator in Riemannian space. The vector u^μ denotes an aether-like field vector, which satisfies the condition $u^\mu u_\mu = \alpha_0(\text{const.})$ in curved spacetime, where α_0 is a real constant. The parameter λ is a small quantity. From a mathematical perspective, $\tilde{\Box} = \text{div}(\text{grad}\,\varphi)$ represents the divergence of the gradient of a scalar field. According to the rules of divergence computation, we obtain

$$\begin{aligned}\tilde{\Box} &= \partial_{;\mu}\left(g^{\mu\nu}\frac{\partial\varphi}{\partial x^\nu}\right) = \frac{\partial}{\partial x^\mu}\left(g^{\mu\nu}\frac{\partial\varphi}{\partial x^\nu}\right) + \Gamma^\nu_{\mu\nu}\left(g^{\mu\nu}\frac{\partial\varphi}{\partial x^\nu}\right)\\ &= \frac{\partial}{\partial x^\mu}\left(g^{\mu\nu}\frac{\partial\varphi}{\partial x^\nu}\right) + \frac{\partial}{\partial x^\mu}(\ln\sqrt{-g})\left(g^{\mu\nu}\frac{\partial\varphi}{\partial x^\nu}\right)\\ &= \frac{1}{\sqrt{-g}}\frac{\partial}{\partial x^\mu}\left(\sqrt{-g}g^{\mu\nu}\frac{\partial\varphi}{\partial x^\nu}\right)\end{aligned} \tag{2}$$

In Equation (2), $\partial_{;\mu}$ denotes the covariant derivative of the contravariant vector field $\left(g^{\mu\nu}\frac{\partial\varphi}{\partial x^\nu}\right) = A^\mu$. Following the conventions of Riemannian geometry, this is computed as $A^\mu_{;\nu} = A^\mu_{,\nu} + \Gamma^\mu_{\lambda\nu}A^\nu$, where $\Gamma^\mu_{\lambda\nu}$ is the Christoffel symbol. Substituting this into the field expression yields Equation (2). In Equation (2), $\Gamma^\nu_{\mu\nu}$ represents the Christoffel symbols associated with the metric tensor of curved spacetime. Based on the variational principle and Equation (1), it can be concluded that the modified bosonic dynamical equations, accounting for Lorentz-breaking theory in curved spacetime, are determined by the following equation:

$$\delta S_b = 0 \tag{3}$$

Noting that $\delta\varphi^2 = 2\varphi\delta\varphi$, $\delta(u^\mu\partial_\mu\phi)^2 = 2(u^\mu\partial_\mu\phi)(u^\nu\partial_\nu\delta\phi)$ and considering that both the integral and variational operators commute, and the variational and differential operators commute, from Equations (1)–(3), the dynamical equation of the scalar field in Lorentz space, modified by Lorentz-breaking theory, is obtained as

$$\frac{1}{\sqrt{-g}}\frac{\partial}{\partial x^\mu}\left(\sqrt{-g}g^{\mu\nu}\frac{\partial\varphi}{\partial x^\nu}\right) + \lambda u^\mu u^\nu\frac{\partial}{\partial x^\mu}\left(\frac{\partial\varphi}{\partial x^\nu}\right) + \frac{m^2}{\hbar^2}\varphi = 0 \tag{4}$$

In fact, this is the modified form of the dynamical equation for a boson with mass m. If we consider the electromagnetic potential A_μ generated by a charged black hole with charge q, then the modified form of the dynamical equation for a boson with mass m and charge e is given by

$$\frac{1}{\sqrt{-g}}\left(D'_\mu\sqrt{-g}g^{\mu\nu}D'_\nu\right)\varphi + \lambda u^\mu u^\nu D'_\mu D'_\nu\varphi + \frac{m^2}{\hbar^2}\varphi = 0 \tag{5}$$

where $D'_\mu = \frac{\partial}{\partial X^\mu} + \frac{ieA_\mu}{\hbar}, D'_\nu = \frac{\partial}{\partial X^\nu} + \frac{ieA_\nu}{\hbar}$. Equations (4) and (5) represent the modified forms of the dynamical equations for spin-zero bosons after incorporating Lorentz-breaking corrections. It is evident that Equation (5) includes an additional correction term compared to the Klein–Gordon equation. We can refer to this equation as the Klein–Gordon–Lorentz equation. In this equation, the correct choice of the aether-like field vector u^μ must be based on the characteristics of the specific black hole spacetime metric tensor $g^{\mu\nu}$. That is, u^μ can be appropriately selected according to the features of different curved spacetimes to ensure the validity of the research results. To illustrate the application of Equation (5) and its associated physical significance, we consider a charged accelerating AdS black hole. According to references [15–18], a charged accelerating AdS black hole is described by the metric and gauge potential

$$ds^2 = \frac{1}{\Omega^2}\left[f(r)dt^2 - \frac{dr^2}{f(r)} - r^2\left(\frac{d\theta^2}{g(\theta)} + g(\theta)\sin^2\theta\frac{d\Phi^2}{K^2}\right)\right] \tag{6}$$

where

$$\begin{aligned} f(r) &= \left(1 - Ar^2\right)\left(1 - \frac{2M}{r} + \frac{q^2}{r^2}\right) + \frac{r^2}{l^2} \\ g(\theta) &= 1 + 2MA\cos\theta + q^2A^2\cos^2\theta \end{aligned} \tag{7}$$

Conformal factor Ω is

$$\Omega = 1 + Ar\cos\theta. \tag{8}$$

The charge q of this black hole and the electromagnetic potential A_t generated by the black hole are given at the event horizon as follows:

$$\Phi = A_t = \frac{q}{r_+} \tag{9}$$

To clarify the acceleration mechanism of the black hole described by Equations (6)–(9), it is essential to examine the underlying process driving this acceleration. For any physical object that interacts with the event horizon of this black hole, its energy–momentum tensor must locally satisfy $T_0^0 \sim T_r^r$. A physical object that meets this condition is a cosmic string. The gravitational effects of a cosmic string induce an overall global conical deficit on a spatial section perpendicular to the string. Hence, the cosmic string essentially represents a localized conical deficit in spacetime. It is precisely due to this conical deficit that the acceleration of the black hole is driven. In the above equations, Ω determines the conformal infinity or boundary of the AdS spacetime. K represents the conical deficit, while M and q correspond to the black hole's mass and electric charge, respectively. $A > 0$ is related to the magnitude of the black hole's acceleration, and $l = \sqrt{\frac{-\Lambda}{3}}$ denotes the AdS radius. One of the notable features of this curved spacetime is that the axis passing through the black hole exhibits a deviation from local flatness. According to Equation (7), the north pole corresponds to $\theta_+ = 0$, while the south pole is at $\theta_- = \pi$. According to Equation (6), the regularity condition of the metric at a pole demands $K_\pm = g(\theta_\pm) = 1 \pm 2MA + q^2A^2$. By fixing, $K_+ = K$ at $\theta = 0$, there is a conical deficit δ on the south pole axis at $\theta = \pi$, given by $\delta = 2\pi\left[1 - \frac{g(\theta_-)}{K_+}\right]$. The parameters in the C-metric described by Equation (6) include the black hole mass M, charge q, acceleration A, cosmological constant l, and tension of the cosmic strings on each axis, encoded by the periodicity of the angular coordinate. For this particular type of black hole, it is necessary to study its thermodynamic properties. From Equation (6), the determinant g of the contra variant metric and the non-zero component of the metric tensor are given as follows:

$$g = -\frac{1}{K^2\Omega^2}r^4\sin^2\theta \tag{10}$$

$$g^{\mu\nu} = \begin{pmatrix} \frac{\Omega^2}{f(r)} & 0 & 0 & 0 \\ 0 & -\Omega^2 f(r) & 0 & 0 \\ 0 & 0 & -\frac{\Omega^2 g(\theta)}{r^2} & 0 \\ 0 & 0 & 0 & -\frac{\Omega^2 K^2}{g(\theta) r^2 \sin^2\theta} \end{pmatrix} \tag{11}$$

The null hypersurface equation describing the black hole horizon is given by

$$g^{\mu\nu} \frac{\partial F}{\partial x^\mu} \frac{\partial F}{\partial x^\nu} = 0 \tag{12}$$

According to Equations (6)–(8), (11) and (12), it follows that the equation that the event horizon of the black hole in question satisfies can be obtained as $g^{rr}\left(\frac{\partial F}{\partial r}\right)^2 = 0$ or $\Omega^2 f(r) = 0$, i.e.,

$$\left(1 + 2Ar\cos\theta + A^2r^2\cos^2\theta\right)\left[\left(1 - Ar^2\right)\left(1 - \frac{2M}{r} + \frac{q^2}{r^2}\right) + \frac{r^2}{l^2}\right] = 0. \tag{13}$$

It can be seen from Equation (13) that $\Omega^2 \neq 0$, and the equation governing the black hole event horizon r_+ is given by

$$f(r_+) = \left(1 - Ar_+^2\right)\left(1 - \frac{2M}{r_+} + \frac{q^2}{r_+^2}\right) + \frac{r_+^2}{l^2} = 0 \tag{14}$$

From Equation (7), it is evident that the acceleration parameter competes with the cosmological constant term $\frac{r^2}{l^2}$ in the Newtonian potential. Alternatively, the negative curvature of the AdS space negates the effect of acceleration. When considering the condition $A < \frac{1}{l}$ in Equations (6)–(8), we describe a single black hole suspended in AdS space, with the only horizon being that of the black hole itself [15–21]. When $A > \frac{1}{l}$, two oppositely charged black holes are present and separated by the acceleration horizon. The case $A = \frac{1}{l}$ corresponds to a special scenario. To ensure the validity of $K_\pm$, the following analysis imposes the constraint $MA < 1/2$, which guarantees that the angular coordinates correspond to those of a standard two-sphere. With the horizon area Equation (14) and its implications established, we proceed to investigate the thermodynamic properties of this black hole. For the purpose of solving Equation (4), we select the aether-like field vector u^μ as follows:

$$u^\mu = \left(\frac{c_t}{\sqrt{g^{tt}}}, \frac{c_r}{\sqrt{g^{rr}}}, \frac{c_\theta}{\sqrt{g^{\theta\theta}}}, \frac{c_\phi}{\sqrt{g^{\phi\phi}}}\right) \tag{15}$$

Equation (15) demonstrates that $u^\mu u_\mu = c_t^2 + c_r^2 + c_\theta^2 + c_\phi^2 = c^2$, where c_t, c_r, c_θ, c_ϕ, and c are constants. Thus, the four components of u^μ selected here satisfy $u^\mu u_\mu = \text{const}$. The WKB approximation, named after Wenzel, Kramers, and Brillouin, is a theoretical method used to solve wave function equations in quantum systems. It constructs the wave function in the form of an exponential function and expands it within the framework of semiclassical theory for solution. According to the WKB semiclassical theory, the wave function φ in Equation (15) can be expressed as

$$\varphi = \varphi_0 e^{i\frac{S}{\hbar}} \tag{16}$$

where S is the action for a boson with electric charge e and mass m. Substituting Equations (15) and (16) into Equation (5) yields

$$(g^{\mu\nu} + \lambda u^\mu u^\nu)\left(\frac{\partial S}{\partial x^\mu} + eA_\mu\right)\left(\frac{\partial S}{\partial x^\nu} + eA_\nu\right) + m^2 = 0 \tag{17}$$

The curved spacetime described by Equation (6) admits a Killing vector ∂_t. Therefore, we obtain

$$\frac{\partial S}{\partial t} = -\omega \tag{18}$$

where ω represents the particle energy. The action S in Equation (17) can be separated into variables as $S = -\omega t + R(r) + Y(\theta) + j\phi$. As can be seen from Equation (9), on the event horizon of this black hole, when $r \to r_+$, $A_t = q/r_+$. Further simplifying Equation (17) yields

$$\begin{aligned}&\frac{\Omega^2}{f(r)}\left(1+\lambda c_t^2\right)\left(\frac{\partial S}{\partial t}+eA_t\right)^2 - \Omega^2 f(r)\left(1+\lambda c_r^2\right)\left(\frac{\partial S}{\partial r}\right)^2 + g^{\theta\theta}\left(1+\lambda c_\theta^2\right)\left(\frac{\partial S}{\partial \theta}\right)^2 \\ &+ g^{\phi\phi}\left(1+\lambda c_\phi^2\right)\left(\frac{\partial S}{\partial \phi}\right)^2 + m^2 = 0\end{aligned} \tag{19}$$

Since $f(r)\big|_{r\to r_+} = 0$, from this equation, we can obtain

$$\left[f(r)\left(\frac{\mathrm{d}R}{\mathrm{d}r}\right)\right]^2_{r\to r_+} = \frac{1+\lambda c_t^2}{1+\lambda c_r^2}\left(\omega - \frac{eq}{r_+}\right)^2 = \frac{1+\lambda c_t^2}{1+\lambda c_r^2}(\omega-\omega_0)^2 \tag{20}$$

or

$$\frac{\mathrm{d}R^\pm}{\mathrm{d}r}|_{r\to r_+} = \pm\left(\frac{1+\lambda c_t^2}{1+\lambda c_r^2}\right)^{\frac{1}{2}} \frac{\omega-\omega_0}{f(r)|_{r\to r_h}}. \tag{21}$$

According to the residue theorem for integration, we obtain

$$R^\pm = \pm i2\pi\left(\frac{1+\lambda c_t^2}{1+\lambda c_r^2}\right)^{\frac{1}{2}} \frac{\omega-\omega_0}{f'(r_+)}, \tag{22}$$

where

$$f'(r_+) = 2\left[\frac{M}{r_+^2} - \frac{q^2}{r_+^3} + AM - Ar_+ + \frac{r_+}{l^2}\right]. \tag{23}$$

According to the quantum tunneling radiation theory, the quantum tunneling rate of a spin-zero, mass m, charge e boson at the event horizon of this black hole is

$$\begin{aligned}\Gamma \sim \exp[-2(\mathrm{Im}S_+ - \mathrm{Im}S_-)] &= \exp[-2(\mathrm{Im}R_+ - \mathrm{Im}R_-)] \\ &= \exp\left[-4\pi\sqrt{\frac{1+\lambda c_t^2}{1+\lambda c_r^2}}\frac{1}{f'(r_+)}(\omega-\omega_0)\right] \\ &= \exp\left(-\frac{\omega-\omega_0}{T_H}\right),\end{aligned} \tag{24}$$

where

$$T_H = \alpha\frac{f'(r_+)}{4\pi} = \alpha\left(\frac{M}{2\pi r_+^2} - \frac{q^2}{2\pi r_+^3} + \frac{AM}{2\pi} - \frac{Ar_+}{2\pi} + \frac{r_+}{2\pi l^2}\right) = \alpha T_h. \tag{25}$$

where $\alpha = \left(\frac{1+\lambda c_t^2}{1+\lambda c_r^2}\right)^{\frac{1}{2}}$. In Equation (20), $\omega_0 = \frac{eq}{r_+}$ is the chemical potential. T_H is the expression of the Hawking temperature at the event horizon of this black hole after being corrected by Lorentz-breaking. Obviously, the quantum tunneling rate of the bosons and the Hawking temperature at the event horizon of a uniformly accelerating charged black hole are not only related to the acceleration parameter A but also to the t-component and r-component of the Lorentz-breaking-related correction term u^μ. When the aether-like correction terms are not considered, $T_H = T_h$, which is completely consistent with the result of the Hawking temperature T_h obtained by the Euclidean method in [15–17].

The Bekenstein–Hawking entropy of this black hole is closely related to its Hawking temperature. If we use the condition $2mA < 1$ to preserve the metric signature, then K encodes information about the conical deficits on the north and south poles with tensions given by [22]

$$u_{\pm} = \frac{\delta_{\pm}}{8\pi} = \frac{1}{4}\left(1 - \frac{1 \pm 2mA + q^2A^2}{K}\right) \tag{26}$$

Therefore, the first law of thermodynamics for this black hole can be expressed as

$$\mathrm{d}M = T_{\hbar}\mathrm{d}S_{bh} + \Phi\mathrm{d}q + V\mathrm{d}P - \lambda_{+}\mathrm{d}u_{+} - \lambda_{-}\mathrm{d}u_{-}, \tag{27}$$

where

$$V = \frac{4}{3}\frac{\pi}{K}\left[\frac{r_+^2}{\left(1 - A^2r_+^2\right)^2}\right], \tag{28}$$

In Equation (27), $\lambda_{\pm}$ denotes the thermodynamic lengths associated with the tension charges of the strings, corresponding to the segments of string attached at each pole. The expression for $\lambda_{\pm}$ is given as follows:

$$\lambda_{\pm} = \frac{r_+}{1 \pm Ar_+} - M\left(1 \pm 2MA + q^2A^2\right). \tag{29}$$

Here, V represents the black hole thermodynamic volume, and P is the pressure associated with the cosmological constant. From Equation (27), we can obtain

$$\begin{aligned} S_{BH} &= \int \frac{\mathrm{d}M - V\,\mathrm{d}p - \phi\mathrm{d}q + \lambda_+\mathrm{d}u_+ + \lambda_-\mathrm{d}u_-}{T_H} \\ &= \left(\frac{1+\lambda c_t^2}{1+\lambda c_r^2}\right)^{-\frac{1}{2}} \int \frac{\mathrm{d}M - V\,\mathrm{d}p - \phi\mathrm{d}q + \lambda_+\mathrm{d}u_+ + \lambda_-\mathrm{d}u_-}{T_h} \\ &= \left(\frac{1+\lambda c_t^2}{1+\lambda c_r^2}\right)^{-\frac{1}{2}} S_{bh}, \end{aligned} \tag{30}$$

In the research on black hole entropy, quantum gravitational effects induce Lorentz symmetry breaking without altering the Bekenstein–Hawking entropy of the black hole. When considering $\hbar$-perturbation theory, logarithmic corrections to the black hole entropy arise. In the present study, we have modified the quantum tunneling rate of this black hole via Lorentz-breaking theory, thereby obtaining the corrected Hawking temperature at the event horizon of the black hole. Consequently, based on the first law of black hole thermodynamics, the modified Bekenstein–Hawking entropy expression as shown in Equation (30) is derived. S_{BH} is the result of the Lorentz-breaking correction to the Bekenstein–Hawking entropy of this black hole, and S_{bh} is the Bekenstein–Hawking entropy of this black hole without considering Lorentz-breaking. Here,

$$S_{bh} = \frac{a_s}{4}. \tag{31}$$

where a_s is the area of the event horizon of this black hole. As can be seen from Equation (6), $\mathrm{d}S' = \frac{1}{\Omega^2}\left(\frac{r^2}{g(\theta)}\mathrm{d}\theta^2 - \frac{g(\theta)r^2\sin^2\theta}{K^2}d\phi^2\right)$. Therefore, we have

$$a_s = \int \mathrm{d}\theta\,\mathrm{d}\phi\sqrt{-g'} = \frac{2\pi r_+^2}{K}\int_0^{\pi} \frac{\sin^2\theta\,\mathrm{d}\theta}{\left(1 + Ar_+\cos\theta\right)^2} = \frac{4\pi r_+^2}{K}\frac{1}{1 - A^2r_+^2}. \tag{32}$$

Substituting Equation (32) into Equation (31), we obtain that the expression of the Bekenstein–Hawking entropy of this black hole before correction is

$$S_{bh} = \frac{\pi r_+^2}{K(1 - A^2 r_+^2)}. \tag{33}$$

Therefore, Equation (30) can be rewritten as

$$S_{BH} = \left(\frac{1+\lambda c_t^2}{1+\lambda c_r^2}\right)^{-\frac{1}{2}} \frac{\pi r_+^2}{K(1 - A^2 r_+^2)}. \tag{34}$$

T_H and S_{BH} are the corrected results of the Hawking temperature and the Bekenstein–Hawking entropy of this black hole obtained within the semi-classical theory. To demonstrate the effect of the quantum perturbation theory on the correction of the Bekenstein–Hawking entropy of this black hole, we first rewrite Equation (21) and its solution Equation (22) as

$$\left.\frac{dR_0^{\pm}}{dr}\right|_{r\to r_+} = \pm\left(\frac{1+\lambda c_t^2}{1+\lambda c_r^2}\right)^{\frac{1}{2}} \frac{E_0}{f(r)|_{r\to r_+}} \tag{35}$$

$$R_0^{\pm} = \pm i\pi\left(\frac{1+\lambda c_t^2}{1+\lambda c_r^2}\right)^{\frac{1}{2}} \frac{E_0}{f'(r_+)}, \tag{36}$$

where $E_0 = \omega - \omega_0$. Considering the quantum perturbation theory, we can express the energy and the radial action of bosons in the spacetime of this black hole as follows:

$$\widetilde{E} = E_0 + \sum_{i=1} \hbar^i E_i \tag{37}$$

$$\widetilde{R}^{\pm} = R_0^{\pm} + \sum_{i=1} \hbar^i R_i^{\pm} \tag{38}$$

Substituting $i = 1, 2, 3, \cdots$ into Equation (20), respectively, we obtain

$$\left.\frac{dR_1^{\pm}}{dr}\right|_{r\to r_+} = \pm\left(\frac{1+\lambda c_t^2}{1+\lambda c_r^2}\right)^{\frac{1}{2}} \frac{E_1}{f(r)|_{r\to r_+}} \tag{39}$$

$$\left.\frac{dR_2^{\pm}}{dr}\right|_{r\to r_+} = \pm\left(\frac{1+\lambda c_t^2}{1+\lambda c_r^2}\right)^{\frac{1}{2}} \frac{E_2}{f(r)|_{r\to r_+}}. \tag{40}$$

Evidently, there exists an inevitable connection between $R_1^{\pm}, R_2^{\pm}, \cdots$ and $R_0^{\pm}$. That is to say, if we denote $\frac{R_i^{\pm}}{R_{i-1}^{\pm}}$ as $\alpha_i' = \frac{\alpha_i}{S_{BH}}$, then after the quantum perturbation correction, the quantum tunneling rate of bosons at the event horizon of this black hole is further modified to

$$\begin{aligned} \widetilde{\Gamma} \sim \exp\left[-2\left(\operatorname{Im}\widetilde{R}_i^{+} - \operatorname{Im}\widetilde{R}_i^{-}\right)\right] &= \exp\left\{-4\pi\left[\left(\frac{1+\lambda c_t^2}{1+\lambda c_r^2}\right)^{\frac{1}{2}}\right]\frac{E_0}{f'(r_+)}\left(1+\sum_{i=1}\hbar^i\alpha_i'\right)\right\} \\ &= \exp\left(-\frac{E_0}{\widetilde{T_H}}\right), \end{aligned} \tag{41}$$

where

$$\widetilde{T_H} = \frac{f'(r_+)}{4\pi}\left(\frac{1+\lambda c_r^2}{1+\lambda c_t^2}\right)^{\frac{1}{2}}\left(1 - \sum_{i=1}\hbar^i\alpha_i'\right). \tag{42}$$

Therefore, when quantum perturbation corrections are considered, the Bekenstein–Hawking entropy of this black hole can be further expressed as

$$\widetilde{S_{BH}} = S_{BH} + \hbar\beta_1 \ln S_{BH} + \cdots \tag{43}$$

where β_1 is the proportionality coefficient between option $\hbar^1$ and $\hbar^0$. This perturbatively corrected result indicates that $\widetilde{S_{BH}}$ has a logarithmic correction [23,24]. The derived black hole entropy demonstrates that it depends not only on the event horizon area but also on the quantum parameter $\hbar^i$ and the coefficients c_t, c_r of the Lorentz-breaking correction terms. These findings highlight the complexity of black hole entropy, which remains a topic warranting further investigation.

3. Lorentz-Breaking Corrections to Quantum Tunneling Radiation of Spin-1/2 Fermions in Accelerating Charged AdS Black Hole Spacetime

In flat spacetime, the action of a spinor field can be expressed in the following two forms [25]:

$$\overline{S} = \int d^4x \overline{\psi'}[i\gamma^m(\partial_m - ieA_m) - m]\psi' \tag{44}$$

$$\overline{S}' = \int d^4x \overline{\psi'}\left[i\gamma^m(\partial_m - ieA_m)\left(1 - \alpha\frac{\gamma^m D_m}{m^2} + b\gamma^5 + \xi(bD)^2\right) - m\right]\psi'. \tag{45}$$

In Equation (45), $D_m = \partial_m - ieA_m$ denotes the standard gauge covariant derivative. Equation (44) represents the action without incorporating Lorentz-breaking corrections, while $\overline{S}'$ corresponds to the action including such corrections. Here, $\overline{\psi'}$ is the complex conjugate of ψ', applicable for spin-1/2 fermions. Since spin-1/2 fermions can have both spin-up and spin-down states, a chiral correction term, $b\gamma^5$, must be introduced into Equation (45) when considering Lorentz-breaking effects. Additionally, the first known Lorentz-breaking term is the Carroll–Field–Jackiw (CFJ) term, which was proposed and explicitly calculated in [26,27]. Therefore, the action of the spinor field must also include the CFJ correction term. In summary, in the curved spacetime of a black hole, considering the Lorentz-breaking theory, we can modify the spinor field action. We propose that the action of the spinor field in curved spacetime be modified to the following form [25]:

$$\begin{aligned} S_f &= \int d^4x\sqrt{-g}\bar{\psi}\left\{i\gamma^\mu D_\mu\left[1 - \frac{a\hbar^2}{m^2}\gamma^\mu\gamma^\nu D_\mu D_\nu\right] + \left(\frac{\lambda}{m}\right)\hbar u^\mu u^\nu D_\mu D_\nu + \left(\frac{b}{m}\right)\frac{\gamma^5}{\hbar} - \frac{m}{\hbar}\right\}\psi \\ &= \int d^4x\sqrt{-g}L_f \end{aligned} \tag{46}$$

where $D_\mu = \partial_\mu + \frac{i}{\hbar}eA_\mu + \Omega_\mu$, $\Omega_\mu = \frac{i}{2}\Gamma_\mu^{\alpha\beta}\Pi_{\alpha\beta}$ and $\Pi_{\alpha\beta} = \frac{i}{4}[\gamma^\alpha, \gamma^\beta]$. In Equation (46), γ^μ denotes the gamma matrices in curved spacetime, which are determined by the specific characteristics of the curved spacetime, and the wave function $\bar{\psi}$ is the conjugate of ψ. The coefficient a represents the CFJ correction term and satisfies the condition $a/m \ll 1$. The coefficient b represents the chiral correction term and satisfies the condition $b/m \ll 1$. The coefficient λ represents the aether-like correction term and satisfies the condition $\lambda/m \ll 1$. The gamma matrices in Equation (46) satisfy the following conditions:

$$\gamma^\mu\gamma^v + \gamma^v\gamma^\mu = 2g^{\mu v}I \tag{47}$$

$$\gamma^\mu\gamma^5 + \gamma^5\gamma^\mu = 0 \tag{48}$$

In the semiclassical WKB theory, the wave function ψ of a spin-1/2 fermion can be expressed as

$$\psi = \begin{pmatrix} \mathrm{C} \\ \mathrm{D} \end{pmatrix} e^{i\frac{S_f}{\hbar}}. \tag{49}$$

where S_f denotes the action of a fermion with charge e, mass m, and spin-1/2. The coefficient of ψ is a 2×1 matrix. According to the variational principle, we can obtain

$$\delta S_f = \int \mathrm{d}^4 x \sqrt{-g}\delta L_f = 0. \tag{50}$$

From Equations (45)–(48) and (51), the modified form of the dynamical equation for a spin-1/2 fermion is obtained as follows:

$$[g^{\mu\nu}(1-2a) + 2\lambda u^\mu u^\nu]\left(\partial_\mu S_f + eA_\mu\right)\left(\partial_\nu S_f + eA_\nu\right) + 2b\gamma_0^5 + m^2 = 0 \tag{51}$$

where γ_0^5 is the coefficient corresponding to γ^5, which must be determined based on the specific characteristics of the curved spacetime. From Equation (6), we can identify the matrix elements of the γ^μ matrices corresponding to the black hole spacetime as follows:

$$\begin{aligned}
\gamma^t &= \sqrt{g^{tt}}\begin{pmatrix} I & 0 \\ 0 & -I \end{pmatrix} = \frac{\Omega}{\sqrt{f(r)}}\begin{pmatrix} I & 0 \\ 0 & -I \end{pmatrix} \\
\gamma^r &= \sqrt{g^{rr}}\begin{pmatrix} 0 & \alpha^1 \\ \alpha^1 & 0 \end{pmatrix} = \Omega\sqrt{-f(r)}\begin{pmatrix} 0 & \alpha^1 \\ \alpha^1 & 0 \end{pmatrix} \\
\gamma^\theta &= \sqrt{g^{\theta\theta}}\begin{pmatrix} 0 & \alpha^2 \\ \alpha^2 & 0 \end{pmatrix} = \frac{\Omega}{r}\sqrt{-g(\theta)}\begin{pmatrix} 0 & \alpha^2 \\ \alpha^2 & 0 \end{pmatrix} \\
\gamma^\phi &= \sqrt{g^{\phi\phi}}\begin{pmatrix} 0 & \alpha^3 \\ \alpha^3 & 0 \end{pmatrix} = \frac{1}{r\sin\theta}\frac{1}{\sqrt{-g(\theta)}}K\Omega\begin{pmatrix} 0 & \alpha^3 \\ \alpha^3 & 0 \end{pmatrix}
\end{aligned} \tag{52}$$

The Pauli matrices in Equation (51) are given by

$$\begin{aligned}
\alpha^1 &= \begin{pmatrix} 0 & 1 \\ 1 & 0 \end{pmatrix} \\
\alpha^2 &= \begin{pmatrix} 0 & -i \\ i & 0 \end{pmatrix} \\
\alpha^3 &= \begin{pmatrix} 1 & 0 \\ 0 & -1 \end{pmatrix}
\end{aligned} \tag{53}$$

Based on Equations (6) and (51), the choice of γ^5 is given by

$$\begin{aligned}
\gamma^5 &= -K\Omega\frac{\sin\theta}{r}\gamma^t\gamma^r\gamma^\theta\gamma^\phi \\
&= -K\Omega\frac{1}{r}\begin{pmatrix} I & 0 \\ 0 & -I \end{pmatrix}\begin{pmatrix} 0 & \alpha^1 \\ \alpha^1 & 0 \end{pmatrix}\begin{pmatrix} 0 & \alpha^2 \\ \alpha^2 & 0 \end{pmatrix}\begin{pmatrix} 0 & \alpha^3 \\ \alpha^3 & 0 \end{pmatrix} \\
&= \gamma_0^5\begin{pmatrix} 1 & 0 \\ 0 & 1 \end{pmatrix}
\end{aligned} \tag{54}$$

where $\gamma_0^5 = K\Omega/r$, and it is clear that γ^5 satisfies Equation (48). The γ^μ and γ^5 chosen from Equations (52)–(54) fully satisfy Equations (47) and (48). Moreover, γ^t, γ^r, γ^θ, γ^ϕ, and γ^5 are all Hermitian matrices, ensuring that the related calculations have physical significance.

By substituting Equations (11), (15) and (54) into the fermion dynamical Equation (51), Equation (51) is specifically given by

$$\frac{\Omega^2}{f(r)}\left(1-2a+2\lambda c_t^2\right)\left(\frac{\partial S_f}{\partial t}+eA_t\right)^2-\Omega^2 f(r)\left(1-2a+2\lambda c_r^2\right)\left(\frac{\partial S_f}{\partial r}\right)^2-\frac{\Omega^2 g(\theta)}{r^2}(1-2a+ \\ 2\lambda c_\theta^2)\left(\frac{\partial S_f}{\partial \theta}\right)^2-\frac{\Omega^2K^2}{g(\theta)r^2\sin^2\theta}\left(1-2a+2\lambda c_\phi^2\right)\left(\frac{\partial S_f}{\partial \phi}\right)^2+\frac{2bK\Omega}{r}+m^2=0 \tag{55}$$

The curved spacetime described by the metric Equation (6) possesses Killing vectors ∂_t and ∂_φ. Therefore, the action in Equation (55) can be separated as

$$S_f=-\omega t+R_f(r)+Y(\theta)+j\phi \tag{56}$$

Substituting Equation (56) into Equation (55) and multiplying both sides of the equation by $f(r)|_{r\to r_H}$, we obtain

$$\left.\frac{dR_f^{\pm}}{\mathrm{d}r}\right|_{r\to r_H}=\pm\frac{1}{f(r)|_{r\to r_H}}\left(\frac{1-2a+2\lambda c_t^2}{1-2a+2\lambda c_r^2}\right)^{1/2}(\omega-\omega_0). \tag{57}$$

Using the residue theorem and setting $R_f^{\pm}=\widetilde{R_0^{\pm}}$, we obtain

$$\widetilde{R_0^{\pm}}=\pm\frac{i2\pi}{f'(r_t)}\left(\frac{1-2a+2\lambda c_t^2}{1-2a+2\lambda c_r^2}\right)^{1/2}(\omega-\omega_0). \tag{58}$$

This yields the quantum tunneling rate of spin-1/2 fermions at the event horizon of a charged, accelerating AdS black hole:

$$\begin{aligned}\Gamma_f\sim\exp\left[-2\left(\mathrm{Im}\,\widetilde{R_0^{+}}-\mathrm{Im}\,\widetilde{R_0^{-}}\right)\right]&=\exp\left[-4\pi\left(\frac{1-2a+2\lambda c_t^2}{1-2a+2\lambda c_r^2}\right)^{1/2}\frac{(\omega-\omega_0)}{f'(r_+)}\right]\\&=\exp\left(-\frac{\omega-\omega_0}{\widetilde{T_H'}}\right)\end{aligned} \tag{59}$$

where $f'(r_+)$ is given by Equation (23). The corresponding expression for $\widetilde{T_H'}$ is provided below:

$$\begin{aligned}\widetilde{T_H'}&=\left(\frac{1-2a+2\lambda c_r^2}{1-2a+2\lambda c_t^2}\right)^{1/2}\left(\frac{M}{2\pi r_+^2}-\frac{q^2}{2\pi r_+^3}+\frac{AM}{2\pi}-\frac{Ar_+}{2\pi}+\frac{r_+}{2\pi l^2}\right)\\&=T_h\left(\frac{1-2a+2\lambda c_r^2}{1-2a+2\lambda c_t^2}\right)^{1/2}\end{aligned} \tag{60}$$

This is the Hawking temperature at the event horizon of the black hole, corrected by Lorentz symmetry breaking. The correction is derived from the modified dynamical equations for spin-1/2 fermions, and it naturally differs from the corresponding result for bosons. The difference primarily manifests in the coefficients of the correction terms. According to the first law of black hole thermodynamics, the black hole entropy is related to the Hawking temperature and other factors. Based on the first law of black hole thermodynamics and $\hbar$ perturbation theory, we can derive the expression for the black hole entropy, which involves terms related to $\hbar^i$, a, c_t, c_r:

$$\widetilde{S_{BH}'}=S_{BH}'+\hbar\ln S_{BH}'+\cdots \tag{61}$$

where

$$S'_{BH} = \left(\frac{1-2a+2\lambda c_t^2}{1-2a+2\lambda c_r^2}\right)^{-1/2} S_{bh} \tag{62}$$

Comparing with Equation (45), it is evident that $\widetilde{S'_{BH}}$ contains an additional coefficient for the CFJ correction term compared to $\widetilde{S_{BH}}$. This correction arises from the spin characteristics of spin-1/2 fermions. It is important to note that the coefficient b of the chiral correction does not affect $\widetilde{S'_{BH}}$ as the equation for $R'(r)$ does not exhibit any b-dependent correction term as $r \to r_+$ due to the condition $f(r_+)b\Omega/r = 0$. However, for $r > r_+$, the fermion energy level distribution in the black hole spacetime will inevitably include corrections that depend on b.

4. Discussion

In the above studies, we considered the case where $A > 0$, which is related to the magnitude of the black hole's acceleration. We applied Lorentz-breaking theories to modify the dynamical equations of bosons and fermions in the spacetime of the black hole. Based on these modifications, we introduced both Lorentz-breaking and quantum perturbation corrections to the Bekenstein–Hawking entropy of the black hole. The results obtained in this work provide valuable insights for studying other types of charged accelerating black holes. As discussed, by setting $M = 0$ and $q = 0$, the black hole horizon is removed, resulting in pure AdS spacetime. This allows for the study of the Rindler horizon for a uniformly accelerating observer. Similarly, the Lorentz-breaking corrections to the Hawking temperature at the Rindler horizon can be explored, along with the correction to the Bekenstein–Hawking entropy. Additionally, we can introduce Rindler-type coordinate transformations in pure AdS spacetime to further investigate the corrections to the Hawking temperature and other physical quantities at the Rindler horizon.The spacetime metric of the black hole is a solution to the Einstein–Maxwell field equation. This solution is a direct generalization of the Reissner–Nordström solution in general relativity and the Bora solution in classical electrodynamics. The results indicate that, except for an additional "outer" Killing horizon due to the accelerated motion, the horizon structure closely resembles that of the Reissner–Nordström case. Kinnersley et al. studied the spacetime characteristics of uniformly accelerating charged black holes and the spacetime metrics of black holes with arbitrary acceleration and no charge [28,29]. Black holes are possibly the most fascinating objects in the study of cosmology and astrophysics.

Author Contributions: Writing—original draft, C.W.; Writing—review & editing, C.W., J.Z. and S.-Z.Y.; Funding acquisition, J.Z. All authors have read and agreed to the published version of the manuscript.

Funding: This work was supported by National Key R&D Program of China (Grant No. 2023YFB4503305) and the National Natural Science Foundation of China (Grant No. 12373109).

Data Availability Statement: Data is contained within the article.

Conflicts of Interest: The authors declare no competing financial interests or personal relationships that may have influenced the work reported in this study.

References

1. 't Hooft, G. On the quantum structure of a black hole. *Nucl. Phys. B* **1985**, *256*, 727–745. [CrossRef]
2. Li, X.; Zhao, Z. Entropy of a Vaidya black hole. *Phys. Rev. D* **2000**, *62*, 104001. [CrossRef]
3. He, F.; Zhao, Z.; Kim, S.W. Statistical entropies of scalar and spinor fields in Vaidya–de Sitter space-time computed by the thin-layer method. *Phys. Rev. D* **2001**, *64*, 044025. [CrossRef]
4. Gao, C.J.; Shen, Y.G. dBB quantization of a torus-like black hole. *Class. Quantum Grav.* **2003**, *20*, 119. [CrossRef]

5. Liu, W.B.; Zhao, Z. The Entropy of Thermal Nonequilibrium Schwarzschild-de Sitter Spacetime. *Acta Math. Sci. Ser. A* **2003**, *23*, 169–174.
6. Dreyer, O. Quasinormal Modes, the Area Spectrum, and Black Hole Entropy. *Phys. Rev. Lett.* **2003**, *90*, 081301. [CrossRef] [PubMed]
7. Nollert, H.P. Quasinormal modes of Schwarzschild black holes: The determination of quasinormal frequencies with very large imaginary parts. *Phys. Rev. D* **1993**, *47*, 5253–5258. [CrossRef] [PubMed]
8. Hod, S. Bohr's Correspondence Principle and the Area Spectrum of Quantum Black Holes. *Phys. Rev. Lett.* **1998**, *81*, 4293. [CrossRef]
9. Amelino-Camelia, G. Phenomenology of Planck-scale Lorentz-symmetry test theories. *New J. Phys.* **2004**, *6*, 188. [CrossRef]
10. Magueijo, J.; Smolin, L. Lorentz Invariance with an Invariant Energy Scale. *Phys. Rev. Lett.* **2002**, *88*, 190403. [CrossRef]
11. Ellis J.; Mavromatos N.E.; Sakharov A.S. Synchrotron radiation from the Crab Nebula discriminates between models of space–time foam. *Astropart. Phys.* **2004**, *20*, 669–682. [CrossRef]
12. Kruglov, S.I. Modified Dirac equation with Lorentz invariance violation and its solutions for particles in an external magnetic field. *Phys. Lett. B* **2012**, *718*, 228–231. [CrossRef]
13. Yang, S.Z.; Lin, K.; Li, J.; Jiang, Q.Q. Lorentz Invariance Violation and Modified Hawking Fermions Tunneling Radiation. *Adv. High Energy Phys.* **2016**, *68*, 190401. [CrossRef]
14. Wang, C.; Tan, X.; Zhang, J.; Li, R.; Yang, S.Z. Correction of Kerr-Sen Black Hole Temperature and Entropy by Lorentz Invariance Violation. *Phys. Scr.* **2024**, *99*, 035009. [CrossRef]
15. Appels, M.; Gregory, R.; Kubizňák, D. Thermodynamics of Accelerating Black Holes. *Phys. Rev. Lett.* **2016**, *117*, 131303. [CrossRef]
16. Krtouš, P. Accelerated black holes in an anti-de Sitter universe. *Phys. Rev. D* **2005**, *72*, 124019. [CrossRef]
17. Gregory, R. Accelerating Black Holes. *arXiv* **2017**, arXiv:1712.04992. [CrossRef]
18. Griffiths, J.B. Podolský, J. A new look at the Plebanski-Demianski family of solutions. *Int. J. Mod. Phys. D* **2006**, *15*, 335–369. [CrossRef]
19. Podolský, J. Accelerating black holes in anti-de Sitter universe. *Czech J. Phys.* **2002**, *52*, 1–10. [CrossRef]
20. Hong, K.; Teo, E. A new form of the C-metric. *Class. Quantum Grav.* **2003**, *20*, 3269. [CrossRef]
21. Emparan, R.; Horowitz, G.T.; Myers, R.C. Exact description of black holes on branes: II. Comparison with BTZ black holes and black strings. *J. High Energy Phys.* **2000**, *2000*, JHEP01(2000). [CrossRef]
22. Appels, M.; Gregory, R.; Kubizňák, D. Black hole thermodynamics with conical defects. *J. High Energy Phys.* **2017**, *2017*, 116. [CrossRef]
23. Shi, Z.J.; Zhang, X.D.; Ma, Y.G. Higher-dimensional quantum Oppenheimer-Snyder model. *Phys. Rev. D* **2024**, *110*, 104074. [CrossRef]
24. Lin, J.H.; Zhang, X.D. Effective four-dimensional loop quantum black hole with a cosmological constant. *Phys. Rev. D* **2024**, *110*, 026002. [CrossRef]
25. Nascimento, J.R.; Petrov, A.Y.; Reyes. C.M. On the Lorentz-breaking theory with higher derivatives in spinor sector. *arXiv* **2015**, arXiv:1505.04968. [CrossRef]
26. Carroll, S.; Field, G.; Jackiw, R. Limits on a Lorentz- and parity-violating modification of electrodynamics. *Phys. Rev. D* **1990**, *41*, 1231. [CrossRef]
27. Jcakiw, R.; Kostelecky, V.A. Radiatively Induced Lorentz and *CPT* Violation in Electrodynamics. *Phys. Rev. Lett.* **1999**, *82*, 3572. [CrossRef]
28. Kinnersley, W.; Walker, M. Uniformly Accelerating Charged Mass in General Relativity. *Phys. Rev. D* **1970**, *2*, 1359. [CrossRef]
29. Kinnersley, W. Field of an Arbitrarily Accelerating Point Mass. *Phys. Rev. D* **1969**, *186*, 1335. [CrossRef]

entropy

MDPI

Article

Correction to Temperature and Bekenstein–Hawking Entropy of Kiselev Black Hole Surrounded by Quintessence

Cong Wang

College of Physics and Electronic Engineering, Qilu Normal University, Jinan 250300, China; wangcong@qlnu.edu.cn

Abstract: This paper studies a rotating Kiselev black hole surrounded by dark energy, whose spacetime metric is a solution to the Einstein field equations. Quintessence is a scalar field with negative pressure, related to the state parameter ω of the dark energy surrounding this black hole. Based on Lorentz-breaking, WKB approximation theory, and quantum tunneling radiation theory, we investigate the characteristic of quantum tunneling radition of spin-1/2 fermions and the result of the correction entropy in this special type of black hole. Additionally, we explore the significance of new expressions for physical quantities such as the Hawking temperature and Bekenstein–Hawking entropy of this black hole.

Keywords: dark energy; quintessence; Kiselev black hole; Lorentz-breaking; Bekenstein–Hawking entropy

1. Introduction

Theoretical and observational studies in astronomy indicate that the universe is currently undergoing accelerated expansion. The dynamical mechanism behind this expansion has not yet been fully determined, leading to the proposal of various models of dark energy. These dark energy models can be broadly categorized into two types. The first type modifies the energy-momentum tensor on the right side of Einstein's field equations, hoping to identify suitable new forms of matter that drive the accelerated expansion of the universe, such as the quintessence scalar field model. The second type aims to amend the mathematical theory of general relativity, which involves modifying the geometric part on the left side of Einstein's field equations, encompassing various modified gravitational theories. Due to significant constraints on the adjustable parameters of these modified gravitational theories, the best model today remains the cosmological constant model, while the concept of new forms of matter is still unclear. The study of dark energy models continues to be a significant topic in contemporary astronomy. The accelerating expansion of the universe implies the contribution of matter with negative pressure to its evolution, which could be the cosmological constant or so-called quintessence matter. If quintessence matter exists throughout the universe, it can also be present around black holes. Therefore, astrophysical black holes are not isolated from matter. Up to now, it is not yet clear what kind of matter dominates the region around black holes. By considering quintessence matter as a source of energy-momentum, researchers have studied the rotating counterpart of the solution to the Einstein equations [1–8]. The rotational Kiselev black hole is surrounded by quintessence matter, and its quantum radiation characteristics represent a topic worthy of in-depth study. The quantum tunneling radiation of black holes can be divided into thermal and non-thermal radiation. Kraus et al. proposed a genuine tunneling radiation theory to investigate the Hawking thermal radiation of black holes [9–15]. Following this, researchers suggested using the semiclassical Hamilton–Jacobi method to study the quantum tunneling radiation of black holes [16–18]. Yang and Lin simplified the Dirac field equation into a matrix equation using semiclassical approximation theory and derived the Hamilton–Jacobi equation in curved spacetime by utilizing the commutation relations of Gamma matrices [19–21]. This method effectively simplifies the related studies of fermionic quantum tunneling radiation. In recent years, research on quantum gravity theory and string theory has indicated that Lorentz dispersion relations require modifications at the Planck scale in high-energy domains [22–25]. Lorentz-breaking can have significant effects on the dynamics of bosons and fermions in curved spacetime, making the study of modifications to curved spacetime and the quantum tunneling radiation characteristics of bosons and fermions a focal point of research [6,7,26–31]. However, in these significant studies, the influence of matter surrounding black holes has not been considered. Therefore, in this paper, we aim to research the impact of dark energy and Lorentz-breaking on the radiations of the Kiselev black hole surrounded by quintessence.

In the next two sections, we introduce the metric characteristics of the static, axially symmetric Kiselev black hole spacetime surrounded by quintessence matter, along with the modified form of the fermionic dynamics equation under Lorentz-breaking theory in this spacetime background. The third section discusses the radiation characteristics of the Kiselev black hole influenced by Lorentz-breaking and dark energy, as well as new expressions and implications for the modified Hawking temperature, Bekenstein–Hawking entropy, and other physical quantities. The final section of this paper discusses the research methods and conclusions derived from this study.

2. Quantum Tunneling Radiation Characteristics of Kiselev Black Hole Surrounded by Quintessence

Black holes in the universe do not exist in isolation. The existence of a black hole is related to the surrounding matter. Therefore, a real black hole spacetime metric should include factors concerning the surrounding matter. Kiselev studied the spacetime metric of a black hole surrounded by quintessence, which is a solution to the Einstein field equations. The Kiselev solution contemplates any type of energy-matter, once a state parameter has been established. The spacetime line element of a rotating Kiselev black hole is given by [6–10]

$$\begin{aligned} ds^2 = & -\left(1-\frac{2Mr+cr^{1-3\omega}}{\Sigma^2}\right)dt^2+\frac{\Sigma^2}{\Delta}dr^2+\Sigma^2 d\theta^2+\sin^2\theta\left(r^2+a^2+a^2\sin^2\theta\frac{2Mr+cr^{1-3\omega}}{\Sigma^2}\right)d\phi^2 \\ & -\frac{2a\sin^2\theta(2Mr+cr^{1-3\omega})}{\Sigma^2}dtd\phi. \end{aligned} \tag{1}$$

where

$$\begin{aligned} \Delta &= r^2-2Mr+a^2-cr^{1-3\omega} \\ \Sigma^2 &= r^2+a^2\cos^2\theta \end{aligned} \tag{2}$$

Here, M is the mass of the black hole, and $a = \frac{J}{M}$ is the angular momentum per unit mass of this rotating black hole. c is the strength parameter. ω is the equation of state (EoS) parameter of the dark energy component. ω defines the EoS, $p = \omega\rho$, where p is the pressure, ρ is the energy density, and for dark energy, the range of ω is

$$-1 < \omega < -\frac{1}{3}. \tag{3}$$

It should be noted that the standard quintessence field is a perfect fluid in the cosmological sense, in the sense that no pressure anisotropy exists. However, in the black hole solution, different values of the quintessence parameter ω correspond to different kinds of energy-matter definable by the EoS. Specifically, dust corresponds to $\omega = 0$, radiation corresponds to $\omega = \frac{1}{3}$, the cosmological constant corresponds to $\omega = 1$, and the so-called R_h universe corresponds to $\omega = -\frac{1}{3}$ [6–10]. Equation (1) describes a black hole wrapped in any kind of energy-matter definable by the EoS. In general, for dark energy, we would expect $\omega < 0$; quintessence, as said, would satisfy Equation (3). From Equations (1) and (2), it can be seen that the non-zero components of $g_{\mu\nu}$ are

$$\begin{aligned} g_{tt} &= -\frac{1}{\Sigma^2}\left(\Sigma^2 - 2Mr - cr^{1-3\omega}\right) \\ g_{rr} &= \frac{\Sigma^2}{\Delta} \\ g_{\theta\theta} &= \Sigma^2 \\ g_{\phi\phi} &= \sin^2\theta\left(r^2 + a^2 + a^2\sin^2\theta\frac{2Mr + cr^{1-3\omega}}{\Sigma^2}\right) \\ g_{t\phi} &= -\frac{a\sin^2\theta(2Mr + cr^{1-3\omega})}{\Sigma^2} \end{aligned} \tag{4}$$

The determinant of the metric corresponding to $g_{\mu\nu}$ obtained from Equation (4) is

$$g = |g_{\mu\nu}| = -\Sigma^4\sin^2\theta = -\left(r^2 + a^2\cos^2\theta\right)^2\sin^2\theta \tag{5}$$

From Equations (4) and (5), as well as the algebraic cofactor $\Delta^{\mu\nu}$ of the metric $g_{\mu\nu}$, the non-zero components of the inverse metric tensor $g^{\mu\nu}$ can be calculated as $g^{\mu\nu} = (-1)^{\mu+\nu}\frac{\Delta^{\mu\nu}}{g}$, and they are, respectively, as follows:

$$\begin{aligned} g^{tt} &= -\frac{1}{\Delta}\left(r^2 + a^2 + a^2\sin^2\theta\frac{2Mr + cr^{1-3\omega}}{\Sigma^2}\right) \\ g^{rr} &= \frac{\Delta}{\Sigma^2} \\ g^{\theta\theta} &= \frac{1}{\Sigma^2} \\ g^{\phi\phi} &= \frac{\Delta - a^2\sin^2\theta}{\Delta\Sigma^2\sin^2\theta} \\ g^{t\phi} &= g^{\phi t} = -\frac{a}{\Delta\Sigma^2}\left(2Mr + cr^{1-3\omega}\right). \end{aligned} \tag{6}$$

In the four-dimensional curved spacetime described by Equation (1), a three-dimensional hypersurface can be represented by the equation $f(t, r, \theta, \phi) = 0$, where r, θ, and ϕ are spherical coordinates. The normal vector is $n_\mu = \frac{\partial f}{\partial x^\mu}$, and the length of the normal vector is given by $n_\mu n^\mu = g^{\mu\nu}\frac{\partial f}{\partial x^\mu}\frac{\partial f}{\partial x^\nu}$. The hypersurface is a null surface if it satisfies the condition $n_\mu n^\nu = 0$. The equation of the zero hypersurface that defines the event horizon of the black hole is as follows

$$g^{\mu\nu}\frac{\partial f}{\partial x^\mu}\frac{\partial f}{\partial x^\nu} = 0 \tag{7}$$

The equation for determining the event horizon of this black hole is $g^{rr} = 0$, obtained by substituting Equation (6) into Equation (7), which gives

$$\Delta = r^2 - 2Mr + a^2 - cr^{1-3\omega} = 0. \tag{8}$$

The horizons of this black hole depend on the equation of state ω and the strength parameter c. When $-1 < \omega < -1/3$, Equation (8) has three horizons corresponding to an event horizon r_+, a Cauchy horizon r_-, and a cosmological horizon. According to

Equation (8), when $\omega \geq -1/3$, the cosmological horizon will disappear. We consider the case $-1 < \omega < -1/3$ and study the quantum tunneling radiation at the black hole's r_H, where we only need to focus on the event horizon. From the equation, we can derive [7,8]

$$r_+ = r_H = M + \sqrt{M^2 - a^2} + \frac{c\left(M + \sqrt{M^2 - a^2}\right)^{1-3\omega}}{2\sqrt{M^2 - a^2}} \tag{9}$$

Equations (1)–(9) describe the spacetime characteristics of the Kiselev black hole surrounded by quintessence. In the curved spacetime of this black hole, the dynamics of bosons and fermions have new expressions after Lorentz-breaking theoretical corrections. Therefore, physical quantities such as the quantum tunneling radiation rate at the event horizon r_H of this black hole will have new results. The following content will present an expression for this topic, which has not been deeply studied. With the expression for r_H, we can study the quantum tunneling radiation characteristics of bosons or fermions at the event horizon of this black hole, as well as related topics. Considering spin-1/2 fermions, which are Dirac particles, and taking into account the Lorentz-breaking correction effects. Nascimento et al. studied the action of the spinor field $\overline{S}$ in flat spacetime and the application of Lorentz-breaking theories, proposing the higher-derivative extension of spinor QED in 4D as follows [26,27]: $\overline{S} = \int d^4x\overline{\varphi}\left[i\gamma^m D_m\left(1 - \alpha\frac{\gamma^m\gamma^n D_m D_n}{m^2}\right) + b\gamma^5 + \xi(bD)^2 - m\right]\varphi$, where $D_m = \alpha m - ieA_m$ is the usual gauge covariant derivative. $\overline{\varphi}$ is the conjugate of φ. α, ξ, b are the coefficients of the CFJ modification term, the aether-like modification term, and the chiral modification term, respectively. The action of the spinor field in the Kiselev black hole spacetime, as expressed by Equations (1) and (2), can be modified to [31]

$$\begin{aligned} S_f = \int \mathrm{d}^4x\sqrt{-g}\widetilde{\psi}\Big\{ & i\gamma^\mu\left(\partial_\mu + \frac{ie}{\hbar}A_\mu + i\Omega_\mu\right)\left[1 - \frac{a_0\hbar^2}{m^2}\gamma^\mu\gamma^\nu\left(\partial_\mu + \frac{ie}{\hbar}A_\mu + i\Omega_\mu\right)\left(\partial_\nu + \frac{ie}{\hbar}A_\nu + i\Omega_\nu\right)\right] \\ & + \frac{\lambda\hbar}{m}u^\mu u^\nu\left(\partial_\mu + \frac{ie}{\hbar}A_\mu + i\Omega_\mu\right)\left(\partial_\nu + \frac{ie}{\hbar}A_\nu + i\Omega_\nu\right) + \frac{b_0\gamma^5}{m\hbar} - \frac{m}{\hbar}\Big\}\psi = \int \mathrm{d}^4x\sqrt{-g}\mathcal{L}_f \end{aligned} \tag{10}$$

According to the variational principle, we can obtain

$$\delta S_f = \int \mathrm{d}^4x\sqrt{-g}\delta\mathcal{L}_f = 0 \tag{11}$$

From Equations (10) and (11), we obtain the modified spin field equation as follows

$$\begin{aligned} \Big\{ & i\gamma^\mu\left(\partial_\mu + \frac{ie}{\hbar}A_\mu + i\Omega_\mu\right)\left[1 - \frac{a_0\hbar^2}{m^2}\gamma^\mu\gamma^\nu\left(\partial_\mu + \frac{ie}{\hbar}A_\mu + i\Omega_\mu\right)\left(\partial_\nu + \frac{ie}{\hbar}A_\nu + i\Omega_\nu\right)\right] \\ & + \frac{\lambda\hbar}{m}u^\mu u^\nu\left(\partial_\mu + \frac{ie}{\hbar}A_\mu + i\Omega_\mu\right)\left(\partial_\nu + \frac{ie}{\hbar}A_\nu + i\Omega_\nu\right) + \frac{b_0\gamma^5}{m\hbar} - \frac{m}{\hbar}\Big\}\psi = 0 \end{aligned} \tag{12}$$

In Equation (10), $\widetilde{\psi}$ is the conjugate of ψ. In Equations (10) and (12), a_0, b_0 and λ are coefficients of small correction terms. Here, a_0 corresponds to the CFJ term, b_0 corresponds to the Chiral correction term, and λ corresponds to the aether-like correction term [26,27,32–34]. According to the WKB approximation theory and black hole quantum tunneling radiation theory, spin-1/2 fermions, also known as Dirac particles, are considered. For Dirac particles, the wave function in Equation (12) can be expressed as

$$\psi = \begin{pmatrix} A \\ B \end{pmatrix} e^{\frac{iS}{\hbar}} \tag{13}$$

Here, A and B are the matrix elements of the 2×1 column matrices. Substituting Equation (13) into Equation (12) and multiplying both sides of Equation (12) by $\hbar$, while omitting the spin connection term $\hbar\Omega_\mu$ that contains $\hbar$, we can obtain the equation for the Dirac particle action S as follows:

$$\left\{-\gamma^\mu\partial_\mu S+\left[1-\frac{a_0}{m^2}\gamma^\mu\gamma^\nu(\partial_\mu S+eA_\mu)(\partial_\nu S+eA_\nu)\right]\left[-\frac{\lambda}{m}u^\mu u^\nu(\partial_\mu S+eA_\mu)(\partial_\nu S+eA_\nu)+\frac{b_0\gamma^5}{m}-m\right]\right\}\begin{pmatrix}A\\B\end{pmatrix}=0 \quad (14)$$

This is a matrix equation, which is actually an eigen matrix equation. The condition of non-trivial solution for this equation to is that the determinant of the corresponding matrix is zero, i.e.,

$$\gamma^\mu\partial_\mu S+\left[1-\frac{a_0}{m^2}\gamma^\mu\gamma^\nu(\partial_\mu S+eA_\mu)(\partial_\nu S+eA_\nu)\right]\left[-\frac{\lambda}{m}u^\mu u^\nu(\partial_\mu S+eA_\mu)(\partial_\nu S+eA_\nu)+\frac{b_0\gamma^5}{m}-m\right]=0. \quad (15)$$

The gamma matrices in Equation (15) are constrained by the following conditions:

$$\gamma^\mu\gamma^\nu+\gamma^\nu\gamma^\mu=2g^{\mu\nu}\mathrm{I} \quad (16)$$

$$\gamma^\mu\gamma^5+\gamma^5\gamma^\mu=0 \quad (17)$$

Thus, it is easy to obtain from Equations (15) and (16) that

$$[g^{\mu\nu}(1-2a_0)+2\lambda u^\mu u^\nu](\partial_\mu S+eA_\mu)(\partial_\nu S+eA_\nu)+m^2-2b_0\gamma_0^5=0. \quad (18)$$

This is the spin field equation for spin-1/2 fermions. In this equation, $g^{\mu\nu}$, u^μ and γ_0^5 need to be determined based on the specific characteristics of the curved spacetime. The conditions $u^\mu u_\mu=$ const., and γ^5 and γ^μ must satisfy Equations (16) and (17). To solve Equation (18), the components of the aether-like field vector u^μ are chosen as follows:

$$\begin{aligned}
u_t&=\frac{c_t}{\sqrt{g^{tt}}}, & u_tu^t&=u_tu_tg^{tt}=c_t^2\\
u_r&=\frac{c_r}{\sqrt{g^{rr}}}, & u_ru^r&=u_ru_rg^{rr}=c_r^2\\
u_\theta&=\frac{c_\theta}{\sqrt{g^{\theta\theta}}}, & u_\theta u^\theta&=u_\theta u_\theta g^{\theta\theta}=c_\theta^2\\
u_\phi&=\frac{c_\phi}{\sqrt{g^{\phi\phi}}}, & u_\phi u^\phi&=u_\phi u_\phi g^{\phi\phi}=c_\phi^2
\end{aligned} \quad (19)$$

Clearly, $u_\mu u^\mu = u_\mu u_\nu g^{\mu\nu} = 2c_t^2+c_r^2+c_\theta^2+2c_\phi^2 = \alpha_0^2(\text{const})$. To facilitate solving Equation (18), we choose the four components of u_μ as shown in Equation (19). In order to determine the specific forms of γ^5 and γ_0^5, we choose the four components of the gamma matrices in the Kiselev spacetime as follows:

$$\begin{aligned}
\gamma^t&=\sqrt{g^{tt}}\begin{pmatrix}\mathbf{I}&0\\0&-\mathbf{I}\end{pmatrix}\\
\gamma^r&=\sqrt{g^{rr}}\begin{pmatrix}0&\sigma^1\\\sigma^1&0\end{pmatrix}\\
\gamma^\theta&=\sqrt{g^{\theta\theta}}\begin{pmatrix}0&\sigma^2\\\sigma^2&0\end{pmatrix}\\
\gamma^\phi&=\frac{g^{t\phi}}{\sqrt{g^{tt}}}\begin{pmatrix}\mathbf{I}&0\\0&-\mathbf{I}\end{pmatrix}+\sqrt{\frac{g^{tt}g^{\phi\phi}-(g^{t\phi})^2}{g^{tt}}}\begin{pmatrix}0&\sigma^3\\\sigma^3&0\end{pmatrix}
\end{aligned} \quad (20)$$

where $\mathbf{I}$ is the identity matrix, and $\sigma^i\,(i=1,2,3)$ are the Pauli matrices, i.e.,

$$\sigma^1=\begin{pmatrix}0&1\\1&0\end{pmatrix},\sigma^2=\begin{pmatrix}0&-i\\i&0\end{pmatrix},\sigma^3=\begin{pmatrix}1&0\\0&1\end{pmatrix} \quad (21)$$

According to the characteristics of the Kiselev spacetime metric, γ^5 is chosen as follows:

$$\begin{aligned}\gamma^5 &= \frac{1}{r^2\sqrt{-\frac{g}{g_{rr}g_{\theta\theta}}}}\gamma^t\gamma^r\gamma^\theta\left[\gamma^\phi - \frac{g^{t\phi}}{\sqrt{g^{tt}}}\begin{pmatrix} \mathbf{I} & 0 \\ 0 & \mathbf{I}\end{pmatrix}\right] \\ &= \frac{\Delta}{i\Sigma^2 r^2}\begin{pmatrix} \mathbf{I} & 0 \\ 0 & -\mathbf{I}\end{pmatrix}\begin{pmatrix} 0 & \sigma^1 \\ \sigma^1 & 0\end{pmatrix}\begin{pmatrix} 0 & \sigma^2 \\ \sigma^2 & 0\end{pmatrix}\begin{pmatrix} 0 & \sigma^3 \\ \sigma^3 & 0\end{pmatrix} = \gamma_0^5\begin{pmatrix} \mathbf{I} & 0 \\ 0 & \mathbf{I}\end{pmatrix}\end{aligned} \tag{22}$$

where

$$\gamma_0^5 = \frac{\Delta}{r^2\Sigma^2} = \frac{r^2 - 2Mr + a^2 - cr^{1-3\omega}}{r^2(r^2 + a^2\cos^2\theta)} \tag{23}$$

Substituting Equations (16), (19) and (21) into the field equation Equation (18), we obtain the dynamical equation for spin-1/2 fermions in the Kiselev spacetime as

$$\begin{aligned}&-\frac{1}{\Delta}\left(1 - 2a_0 + 2\lambda c_t^2\right)\left[\Sigma^2\left(r^2 + a^2\right) + a^2\left(2Mr + cr^{1-3\omega}\right) - a^2\cos^2\theta\left(2Mr + cr^{1-3\omega}\right)\right]\left(\frac{\partial S}{\partial t}\right)^2 \\ &+\Delta\left(1 - 2a_0 + 2\lambda c_r^2\right)\left(\frac{\partial S}{\partial r}\right)^2 - \frac{2}{\Delta}\left(1 - 2a_0 + 2\lambda c_\phi^2\right)a\Sigma^2\left(2Mr + cr^{1-3\omega}\right)\left(\frac{\partial S}{\partial t}\right)\left(\frac{\partial S}{\partial \phi}\right) \\ &+\frac{1}{\Delta}\left(1 - 2a_0 + 2\lambda c_\phi^2\right)\left(\frac{\Delta}{\sin^2\theta} - a^2\right)\left(\frac{\partial S}{\partial \phi}\right)^2 + \Sigma^2\left[m^2 - 2b_0\gamma_0^5 + g_{\theta\theta}\left(1 - 2a_0 + 2\lambda c_\theta^2\right)\right]\left(\frac{\partial S}{\partial \theta}\right)^2 \\ &+\Sigma^2\left(m^2 - 2b_0 r_0^5\right) + \lambda u^r\left(u^t\frac{\partial S}{\partial r}\frac{\partial S}{\partial t} + u^\theta\frac{\partial S}{\partial r}\frac{\partial S}{\partial \theta} + u^\phi\frac{\partial S}{\partial r}\frac{\partial S}{\partial \phi}\right) + 4\lambda u^\theta\left(u^t\frac{\partial S}{\partial \theta}\frac{\partial S}{\partial t} + u^\phi\frac{\partial S}{\partial \theta}\frac{\partial S}{\partial \phi}\right)^2 = 0\end{aligned} \tag{24}$$

where $\Sigma^2 = r^2 + a^2\cos^2\theta = r^2 + a^2 - a^2\sin^2\theta$. The constant introduced when separating variables in Equation (24) is denoted as d_0. Multiplying both sides of this equation by Δ, and noting that $\Delta|_{r\to r_H} = 0$, the radial motion equation for spin-1/2 particle, derived from Equation (24) at the black hole horizon, is given by

$$\begin{aligned}&-\left(1 - 2a_0 + 2\lambda c_t^2\right)\left[\left(r_H^2 + a^2\right)r_H^2 + a^2\left(2Mr_H + cr_H^{1-3\omega}\right)\right]\left(\frac{\partial S}{\partial t}\right)^2 \\ &+\left(1 - 2a_0 + 2\lambda c_r^2\right)\left(\Delta\frac{\partial S}{\partial r}\right)^2_{r\to r_H} + 2\left(1 - 2a_0 + 2\lambda c_\phi^2\right)a\left(2Mr_H + cr_H^{1-3\omega}\right)\frac{\partial S}{\partial t}\frac{\partial S}{\partial \phi} \\ &-\left(1 - 2a_0 + 2\lambda c_\phi^2\right)a^2\left(\frac{\partial S}{\partial \phi}\right)^2 + \left[\Delta|_{r\to r_H}\left(m^2 - 2b_0\gamma_0^5\right)r - \Delta d_0\right]_{r\to r_H} = 0\end{aligned} \tag{25}$$

The Kiselev spacetime possesses a Killing vector $\frac{\partial}{\partial \phi}$, so $\frac{\partial S}{\partial \phi} = \mathrm{n}(\mathrm{const.})$ the particle action S in Equation (25) can be separated, i.e.,

$$S = -Et + R(r) + \Theta(\theta) + \mathrm{n}\phi \tag{26}$$

After substituting Equation (26) into Equation (25), the following is obtained:

$$\begin{aligned}\left(\Delta\frac{\mathrm{d}R}{\mathrm{d}r}\right)^2_{r\to r_H} &= \frac{1 - 2a_0 + 2\lambda c_t^2}{1 - 2a_0 + 2\lambda c_r^2}\left\{\left[E\left(r_H^2 + a^2\right)\right]^2 - 2a\mathrm{n}E\left(2Mr_H + cr_H^{1-3\omega}\right) + a^2\mathrm{n}^2\right\} \\ &= \frac{1 - 2a_0 + 2\lambda c_t^2}{1 - 2a_0 + 2\lambda c_r^2}\left(r_H^2 + a^2\right)^2(E - E_0)^2\end{aligned} \tag{27}$$

Got Equation (27) before we used $u^t u^\phi = g^{t\phi}u_\phi u^\phi = g^{t\phi}c_\phi^2$ and $u^t u^\phi = u^t u_t g^{t\phi} = g^{t\phi}c_t^2$. From Equation (27), we obtain

$$E_0 = \frac{a\mathrm{n}}{r_H^2 + a^2} \tag{28}$$

$$\left.\frac{dR}{dr}\right|_{r\to r_H}=\pm\frac{1}{\Delta|_{r\to r_H}}\left(\frac{1-2a_0+2\lambda c_t^2}{1-2a_0+2\lambda c_r^2}\right)^{1/2}\left(r_H^2+a^2\right)(E-E_0). \tag{29}$$

The application of the residue theorem to integrate both sides of Equation (28) yields

$$R_\pm=\pm\frac{i\pi(r_H^2+a^2)}{\Delta'(r_H)}\left(\frac{1-2a_0+2\lambda c_t^2}{1-2a_0+2\lambda c_r^2}\right)^{1/2}(E-E_0), \tag{30}$$

where

$$\Delta'(r_H)=2r_H-2M-c(1-3\omega)r_H^{-3\omega}. \tag{31}$$

Based on the WKB approximation and the quantum tunneling radiation theory, the quantum tunneling rate of the semiclassical at the event horizon r_H of the Kiselev black hole is given by

$$\Gamma\sim\exp(-2\,\mathrm{Im}\,S_\pm)=\exp(-2\,\mathrm{Im}\,R_\pm)=\exp\left(-\frac{E-E_0}{T_H}\right), \tag{32}$$

where

$$T_H=\frac{2r_H-2M+c(1-3\omega)r_H^{1-3\omega}}{4\pi(r_H^2+a^2)}\left(\frac{1-2a_0+2\lambda c_r^2}{1-2a_0+2\lambda c_t^2}\right)^{1/2}. \tag{33}$$

To elucidate the relationship between the corrected entropy and the $\hbar$-perturbation, Equations (29) and (30) can be, respectively, rewritten in the following forms:

$$\left.\frac{dR_0^\pm}{dr}\right|_{r\to r_H}=\pm\frac{1}{\Delta|_{r\to r_H}}\left(\frac{1-2a_0+2\lambda c_t^2}{1-2a_0+2\lambda c_r^2}\right)^{1/2}\left(r_H^2+a^2\right)(E-E_0) \tag{34}$$

$$R_0^\pm=\pm\frac{i\pi(r_H^2+a^2)}{\Delta'(r_H)}\left(\frac{1-2a_0+2\lambda c_t^2}{1-2a_0+2\lambda c_r^2}\right)^{1/2}(E-E_0) \tag{35}$$

Let

$$E=E_0'+\sum_{i=1}\hbar^iE_i \tag{36}$$

$$R^\pm=R_0^\pm+\sum_{i=1}\hbar^iR_i^\pm \tag{37}$$

where $E_0'=E-E_0$. From Equation (29) to Equation (37), the following relationship can be derived [35,36]

$$\left.\frac{dR_1^\pm}{dr}\right|_{r\to r_H}=\pm\frac{E_1(r_H^2+a^2)}{\Delta|_{r_1\to r_H}}\left(\frac{1-2a_0+2\lambda c_t^2}{1-2a_0+2\lambda c_r^2}\right)^{1/2} \tag{38}$$

$$\left.\frac{dR_2^\pm}{dr}\right|_{r\to r_H}=\pm\frac{E_2(r_H^2+a^2)}{\Delta|_{r_1\to r_H}}\left(\frac{1-2a_0+2\lambda c_t^2}{1-2a_0+2\lambda c_r^2}\right)^{1/2}$$
$$\vdots \tag{39}$$

From Equations (9), (38) and (39), it is clear that an inherent relationship exists between $R_1^\pm$, $R_2^\pm$, $R_3^\pm\cdots$ and $R_0^\pm$. Defining $\frac{R_i^\pm}{R_{i+1}^\pm}=\alpha_i'$, Given $E_i/E_{i-1}=\beta_i$, Equations (36) and (37) can be rewritten as $E=E_0'+\sum\limits_{i=1}\hbar^i\beta_iE_0'$, $R^\pm=R_0^\pm+\sum\limits_{i=1}\hbar^i\alpha_i'R_0^\pm$. where α_i' can be expressed as α_i/S_{BH}. T_H' denotes the Hawking temperature at the event horizon of this black hole, which includes terms with $\hbar$ corrections. According to Equation (37), the quantum tunneling rate of this black hole can be rewritten as

$$\begin{aligned}\Gamma' \sim \exp(-2I_m R_i^{\pm}) &= \exp\left[-4\pi\frac{(r_H^2+a^2)E_0'}{\Delta'(r_H)}\left(\frac{1-2a_0+2\lambda c_t^2}{1-2a_0+2\lambda c_r^2}\right)^{1/2}\left(1+\sum_{i=1}\hbar^i\alpha_i'\right)\right] \\ &= \exp\left(-\frac{E_0'}{T_H'}\right)\end{aligned} \tag{40}$$

where

$$T_H' = \frac{\Delta'(r_H)}{4\pi(r_H^2+a^2)}\left(\frac{1-2a_0+2\lambda c_t^2}{1-2a_0+2\lambda c_r^2}\right)^{-1/2}\left(1+\sum_{i=1}\hbar^i\alpha_i'\right)^{-1} = T_H\left(1+\sum_{i=1}\hbar^i\alpha_i/S_{BH}\right)^{-1} \tag{41}$$

For this black hole, the corrected first law of thermodynamics is formulated as follows:

$$\mathrm{d}M = T_H'\mathrm{d}S_{BH}' + \Omega\mathrm{d}J - \rho\mathrm{d}V_d \tag{42}$$

Thus,

$$\begin{aligned}S_{BH}' &= \int\frac{\mathrm{d}M - \Omega\mathrm{d}J - \rho\mathrm{d}V_d}{T_H'} \\ &= S_{BH} + \hbar\alpha_1\ln S_{BH} + \cdots\end{aligned} \tag{43}$$

where

$$\begin{aligned}S_{BH} &= \int\frac{\mathrm{d}M - \Omega\mathrm{d}J - e\mathrm{d}V_d}{T_H} = \int\frac{\mathrm{d}M - \Omega\mathrm{d}J - e\mathrm{d}V_d}{T_0}\left(\frac{1-2a_0+2\lambda c_t^2}{1-2a_0+2\lambda c_r^2}\right)^{1/2} \\ &= S_{bh}\left(\frac{1-2a_0+2\lambda c_t^2}{1-2a_0+2\lambda c_r^2}\right)^{1/2}\end{aligned} \tag{44}$$

where $T_0 = \frac{\Delta'(r_H)}{4\pi(r_H^2+c_r^2)}$ is the original temperature of the black hole, and S_{bh} is the original Bekensein–Hawking entropy. Equation (43) is the logarithmic-corrected version of the Bekenstein–Hawking entropy of this black hole. The logarithmic-corrected version of the Bekenstein–Hawking entropy is approved by various quantum efforts such as conical singularity and entanglement entropy, Euclidean action method, quantum geometry, Candy formula, conformal anomaly, Noether charge, and non-locality of quantum gravity [35,36]. According to the new expression for the Bekenstein–Hawking entropy of the Kiselev black hole and the second law of black hole thermodynamics, the black hole entropy never decreases in the clockwise direction, i.e.,

$$\Delta S_{BH}' \geq 0 \tag{45}$$

Therefore, the expression for the black hole's quantum tunneling radiation rate in Equation (40) can be rewritten as

$$\Gamma = \exp(\Delta S_{BH}'). \tag{46}$$

The area of the black hole event horizon is A_k . Due to

$$\mathrm{d}H^2 = \Sigma^2 d\theta^2 + \sin^2\theta\left(r_H^2 + a^2 + a^2\sin^2\theta\frac{2Mr_H + cr_H^{1-3\omega}}{\Sigma^2}\right)\mathrm{d}\phi^2, \tag{47}$$

the metric determinant corresponding to Equation (47) is

$$g_H = \sin^2\theta\left(r_H^2 + a^2\right)\left(r_H^2 + a^2\cos^2\theta\right) + a^2\sin^2\theta\left(r_H^2 + a^2\right) = \sin^2\theta\left(r_H^2 + a^2\right)^2 \tag{48}$$

where r_H is related to c and ω .

$$A_k = \int \sqrt{g_H}\mathrm{d}\theta\mathrm{d}\phi = 4\pi\left(r_H^2 + a^2\right) = 4\pi\left(2Mr_H + cr_H^{1-3\omega}\right) \tag{49}$$

so,

$$S_{bh} = A_k/4 \tag{50}$$

Thus, the semiclassical Bekenstein–Hawking area law is given by $S_{bh} = \frac{A_k}{4} = \pi(2Mr_H + cr_H^{1-3\omega})^2$ represents the uncorrected original Bekenstein–Hawking entropy of the black hole and A_k is the area of the event horizon of the black hole. S_{BH} is corrected semiclassical Bekenstein–Hawking entropy. S'_{BH} is a type of the Bekenstein–Hawking entropy included logarithmic-corrected. Through S_{bh}, S_{BH}, and S'_{BH}, our understanding of the entropy of the black hole has deepened. Equation (32) represents a new expression for the quantum tunneling rate of spin-1/2 fermions at the event horizon r_H of the Kiselev black hole, where T_H denotes the Hawking temperature at the black hole horizon. Given that $\Delta|_{r\to r_H} = 0$, we derive the relation $r_H^2 + a^2 = 2Mr_H + cr_H^{1-3\omega}$. The expressions for Γ, T_H, and ω_0 demonstrate that these quantities are all dependent on $r_H^2 + a^2$, which in turn depends on c and ω. Therefore, the quantum tunneling radiation of the Kiselev black hole surrounded by quintessence matter is characterized by distinctive features. Modifications such as quintessence matter, CFJ corrections, and aether-like field modifications each contribute to the quantum tunneling radiation of the black hole. The coefficient of the chiral correction term, denoted b_0, does not influence the quantum tunneling rate of the black hole, thus having no impact on the thermal quantum radiation described above.

3. Research on the Impact of Dark Energy and Lorentz-Breaking on Dirac Energy Levels in the Spacetime of the Kiselev Black Hole Surrounded by Quintessence

When examining quantum non-thermal radiation, the chiral correction term affects the energy level distribution of Dirac particles in the black hole spacetime, as well as the maximal crossing between the positive and negative energy levels of Dirac particles near the event horizon. To investigate the impact of dark energy and Lorentz-breaking on the Dirac energy levels in the curved spacetime of the black hole, we consider the equation governing the radial action R of Dirac particles in the region where $r > r_H$ within the curved spacetime, as shown below:

$$\left(\Delta\frac{\mathrm{d}R}{\mathrm{d}r}\right)^2 = \frac{1-2a_0+2\lambda c_t^2}{1-2a_0+2\lambda c_r^2}\left\{\left[E\left(r_H^2+a^2\right)\right]^2 - 2an E\left(2Mr_H + cr_H^{1-3\omega}\right) + a^2 n^2 - \Delta\frac{(m^2-2b_0\gamma_0^5)r^2 - d_0}{1-2a_0+2\lambda c_r^2}\right\} \tag{51}$$

The left-hand side of this equation is a positive real number, thereby necessitating that the right-hand side satisfies the following conditions:

$$\left[E^2\left(r_H^2+a^2\right)\right]^2 - 2anE\left(2Mr_H + cr_H^{1-3\omega}\right) + a^2 n^2 - \Delta\frac{(m^2-2b_0\gamma_0^5)r^2 - d_0}{1-2a_0+2\lambda c_r^2} \geq 0 \tag{52}$$

By solving for the equality, we obtain the equation satisfied by the positive and negative energy levels of Dirac particles in the spacetime of the Kiselev black hole as follows:

$$\left[E\left(r_H^2+a^2\right) - an\right]^2 = \Delta\frac{(m^2-2b_0\gamma_0^5)r^2 - d_0}{1-2a_0+2\lambda c_t^2} \tag{53}$$

From this equation, the positive and negative energy levels of Dirac particles in the spacetime of a Kiselev black hole are given by the following:

$$E_{\pm} = \frac{an}{r_H^2 + a^2} \sqrt{\frac{\Delta}{\left(r_H^2 + a^2\right)^2} \frac{\left(m^2 - 2b_0\gamma_0^5\right)r^2 - d_0}{1 - 2a_0 + 2\lambda c_r^2}} \tag{54}$$

It follows that the parameters $a_0, \lambda, b_0, c, \omega$ all induce certain modifications to the distribution of positive and negative energy levels of the particles. From Equation (54), the energy level distribution in the spacetime of this black hole is given by

$$\begin{aligned} E &> E_+ \\ E &< E_- \end{aligned} \tag{55}$$

As $r \to r_H$, Equation (55) reveals that

$$\lim_{r \to r_H} E_{\pm} = \frac{an\left(2Mr_H + cr_H^{1-3\omega}\right)}{\left(r_H^2 + a^2\right)^2} = \frac{an}{r_H^2 + a^2} = E_0 \tag{56}$$

where E_0 represents the maximum value of the crossover energy levels E_+ and E_-. Under these conditions, the maximum energy condition for the quantum non-thermal radiation particles is $E \leq E_0$. Considering the limit $r \to \infty$, Equation (50) reveals that

$$E|_{r \to \infty} \sim \pm m \tag{57}$$

where m denotes the mass of the spin-1/2 fermion. By combining Equations (56) and (57), the energy range of quantum non-thermal radiation at the black hole's event horizon r_H is given by

$$-m < E \leq E_0 \tag{58}$$

From Equations (8), (9) and (55)–(58), it can be seen that the distribution of fermionic levels in the spacetime of this black hole surrounded by dark energy is related to the dark energy component ω. This is a significant difference from other stationary black holes. Equation (58) can be considered as a new expression for the energy range of quantum non-thermal radiation particles from stationary black holes. The physical significance of this expression is as follows: particles exhibit wave-particle duality, and radiation particles can exist in both positive and negative energy states. There is a forbidden zone between these energy states, where no particles can exist. When the energy E in the negative energy state satisfies the condition in Equation (58), a phenomenon similar to the Klein mechanism occurs, where particle pairs are generated through quantum tunneling, resulting in radiation particles. This effect is known as quantum non-thermal radiation. In fact, this is a form of stimulated radiation of the black hole. When a black hole undergoes stimulated radiation, spontaneous radiation is also inevitably present, which corresponds to the Starobinsky–Unruh process. Both stimulated and spontaneous radiation are independent of the black hole's temperature, and the radiative quantum non-thermal effects cause a reduction in the black hole's rotation and angular momentum.

4. Discussion

The quantum tunneling rate Γ', Hawking temperature T'_H, and Bekenstein–Hawking entropy S'_{BH} of this black hole after Lorentz-breaking and quantum perturbative corrections are given by Equations (40), (41) and (43), respectively.

Equations (32) and (33) are the corresponding results in semiclassical theory. By denoting the uncorrected Hawking temperature and tunneling rate as T_0 and Γ_0, respectively, the ratio of the tunneling rates before and after the correction can be obtained as follows:

$$\frac{\Gamma_0}{\Gamma} = \exp\left\{\left(-\frac{E-E_0}{T_0}\right)\left[1-\left(\frac{1-2a_0+2\lambda c_t^2}{1-2a_0+2\lambda c_r^2}\right)^{1/2}\right]\right\} \tag{59}$$

Equations (40) and (41) represent the corresponding results including $\hbar$ perturbation corrections. Therefore,

$$\frac{\Gamma_0}{\Gamma'} = \exp\left\{\left(-\frac{E-E_0}{T_0}\right)\left[1-\left(\frac{1-2a_0+2\lambda c_t^2}{1-2a_0+2\lambda c_r^2}\right)^{1/2}\left(1+\sum_{i=1}\hbar^i\alpha_i'\right)\right]\right\}. \tag{60}$$

Equations (59) and (60) show the effects of Lorentz-breaking corrections and $\hbar$ perturbation corrections on the quantum tunneling rate of this black hole. Since $T_H(T_H')$ is related to the dark energy parameter ω, and T_0, Γ, Γ', and the Bekenstein–Hawking entropy are all related to the Hawking temperature of the black hole, it is necessary to further explain the effect of the dark energy parameter ω on the black hole temperature and the Bekenstein–Hawking entropy. Using $T_H = \frac{2r_H-2M+c(1-3\omega)r_H^{1-3\omega}}{4\pi(r_H^2+a^2)}\left(\frac{1-2a_0+2\lambda c_t^2}{1-2a_0+2\lambda c_r^2}\right)^{1/2} = T_0\left(\frac{1-2a_0+2\lambda c_t^2}{1-2a_0+2\lambda c_r^2}\right)^{1/2}$ and $S_{BH} = S_{bh}\left(\frac{1-2a_0+2\lambda c_t^2}{1-2a_0+2\lambda c_r^2}\right)$ as an example, the dependence of the temperature and the entropy and other physical quantities on ω is shown in the form of a graph.

From this table, it can be seen that when ω takes values in the range $-1 < \omega < -\frac{1}{3}$, as ω increases, both T_0 and T_H decrease. This reflects the influence of dark energy on the Hawking temperature of this black hole. Considering the Lorentz-breaking correction effects, the same value of ω corresponds to different T_0 and T_H, with $T_H < T_0$, which indicates that the Lorentz-breaking effect modifies T_0. This is what distinguishes this black hole from other types of black holes and makes it unique. From Table 1, it can be seen that ω has a certain effect on S_{bh}. S_{BH} is influenced not only by ω but also by the Lorentz-breaking correction parameter. As ω increases, both S_{bh} and S_{BH} decrease.

Table 1. The effect of the dark energy parameter ω on the black hole temperature.

ω	**−0.85**	**−0.8**	**−0.75**	**−0.7**	**−0.6**	**−0.5**	**−0.4**	**−0.37**	**−0.34**
T_0	0.908426	0.675019	0.510699	0.393292	0.245706	0.164148	0.116817	0.106711	0.0979722
T_H	0.900012	0.668767	0.505969	0.389649	0.24343	0.162628	0.115735	0.105722	0.0970647
S_{bh}	169.612	124.936	94.8502	74.1562	49.2101	35.9299	28.3088	26.673	25.2503
S_{BH}	166.484	122.632	93.1013	72.7889	48.3027	35.2674	27.7869	26.1812	24.7847

$M = 1, a = 0.6, c = 0.4, a_0 = 0.1, \lambda = 0.11, c_t = 0.3, c_r = 0.4$.

We can use the Stefan–Boltzmann law to estimate the radiation rate and lifetime of this black hole. This helps illustrate the thermodynamic evolution of the black hole. According to the Stefan–Boltzmann law, the radiation rate of this black hole is as follows:

$$\frac{\mathrm{d}E'}{\mathrm{d}t} = \sigma' A_K (T_H')^4 \tag{61}$$

where σ' is the Stefan–Boltzmann constant. A_k is given by Equation (48), and T_H' is given by Equation (41). Clearly, M, ω, a_0, c_r, c_t, and $\hbar^i$ all influence $\frac{\mathrm{d}E'}{\mathrm{d}t}$. E' is the total energy of this black hole. By substituting Equations (41) and (33) into Equation (61), we obtain the following:

$$\begin{aligned}\frac{dE'}{dt} &= \left(\frac{dE'}{dt}\right)_L \left(1 + \sum_{i=1} \hbar^i \alpha_i / S_{BH}\right)^{-4} \\ &= \left(\frac{dE'}{dt}\right)_o \left(\frac{1 - 2a_0 + 2\lambda c_r^2}{1 - 2a_0 + 2\lambda c_t^2}\right)^{-2} \left(1 + \sum_{i=1} \hbar^i \alpha_i / S_{BH}\right)^{-4}\end{aligned} \tag{62}$$

where $\left(\frac{dE'}{dt}\right)_L$ is the radiation rate that includes the Lorentz-breaking corrections. $\left(\frac{dE'}{dt}\right)_o$ is the original radiation rate without corrections. From Equation (62), it can be seen that Lorentz-breaking and $\hbar^i$ perturbative corrections have certain effects on the thermodynamic evolution of this black hole. When the mass of the black hole reduces to the minimum value $M_{\min}$, the black hole will no longer radiate particles and will leave a residual mass $M_{\text{res}} = M_{\min}$ [37]. We can estimate the evolution lifetime of this black hole. In Equations (61) and (62), $\sigma' = \frac{2\pi^5 K_B^4}{15\hbar^3 c^2}$ and $A_k \sim 16\pi G^2 c^{-4} M^2$, therefore, the radiation rate of this black hole is estimated as follows:

$$\frac{dE'}{dt} \sim \left(\frac{dE'}{dt}\right)_o \Gamma' \sim \left(\frac{dE'}{dt}\right)_o \Gamma(erg/s) \tag{63}$$

where M is in grams. The tunneling rates Γ' and Γ are both estimated to be 1. It can be calculated that the radiation rate of this black hole is approximately:

$$\frac{dE'}{dt} \sim 10^{46}(M^{-2})(erg/s) \tag{64}$$

Therefore, the lifetime of this black hole is estimated as $t \sim 10^{-27} M^3 S \sim 10^{67}\left(\frac{M}{M_\odot}\right)^3$ (year). This shows that the lifetime of the black hole is directly related to its mass. The larger the mass M, the longer the lifetime t, and vice versa.The main content of the above research is the thermodynamic evolution characteristics of this black hole under the modifications of Lorentz-breaking and $\hbar$ perturbation. Based on the modifications of the Hawking temperature and Bekenstein–Hawking entropy under Lorentz-breaking and $\hbar$ perturbation theory, it can be seen that the specific heat capacity of this black hole, $C = T'_H\left(\frac{\partial S'_{BH}}{\partial T'_H}\right)$, is also modified. These modifications are all meaningful research results. Since the dynamical Equation (18) for spin-1/2 fermions with Lorentz-breaking corrections is a semiclassical equation, the quantum tunneling rate, Hawking temperature and Bekenstein–Hawking entropy of the black hole obtained above are all results within the semiclassical theory, and these results are related to ω. It should be noted that the range of ω considered in the above study is $-1 < \omega < -\frac{1}{3}$. When $\omega = 0$, it corresponds to the dust case, while $\omega = -\frac{1}{3}$ represents the radiation case. $\omega = -1$ corresponds to the cosmological constant model, and $\omega = -\frac{1}{3}$ is associated with the so-called R_h universe model [38,39]. Additionally, it is important to emphasize that the results presented above are derived within the framework of semiclassical approximation theory. To incorporate the quantum corrections due to the Planck constant $\hbar$, the term $\hbar\Omega_u$ should be retained in Equation (18). This modification enables the study of perturbative effects of $\hbar$ based on the semiclassical Equation (18). The classical components of $R_\pm$ and E can be expressed as a series expansion in powers of $\hbar$, yielding expressions for quantum tunneling, Hawking temperature, and Bekenstein–Hawking entropy with higher-order corrections in $\hbar$. Furthermore, it is essential to highlight that all types of black holes exhibit quantum thermal effects, which are linked to the Hawking temperature. These effects are present in both stationary and non-stationary black holes, which exhibit both quantum thermal effects and quantum non-thermal effects. Static black holes, however, do not display quantum non-thermal effects. It should be further explained that we chose Equation (19) and defined u_μ instead of u^μ. Our choice can simplify the calculation. The above research methods and conclusions are meaningful for the study of dynamical corrections of fermions with arbitrary spins and the tunneling of other types of black holes. The methods employed in this study are not applicable to bosons, and the relevant theory for scalar fields should be used for further investigation. It should be further noted that the above research methods can be applied to study quantum tunneling radiation specific characteristics of other fermions.

Funding: This work was supported by National Key R&D Program of China (Grant No. 2023YFB4503305) and the National Natural Science Foundation of China (Grant No. 12373109).

Data Availability Statement: Data is contained within the article.

Acknowledgments: The author would like to express gratitude to Shu-Zheng Yang for fruitful discussions and valuable insights.

Conflicts of Interest: The author declares no competing financial interests or personal relationships that may have influenced the work reported in this study.

References

1. Ghosh, S.G. Rotating black hole and quintessence. *Eur. Phys. J. C* **2016**, *76*, 222. [CrossRef]
2. Kiselev, V.V. Quintessence and black holes. *Class. Quantum Gravity* **2003**, *20*, 1187. [CrossRef]
3. Thomas, B.B.; Saleh, M.; Kofane, T.C. Thermodynamics and phase transition of the Reissner–Nordström black hole surrounded by quintessence. *Gen. Relativ. Gravit.* **2012**, *44*, 2181–2189. [CrossRef]
4. Fernando, S. Nariai black holes with quintessence. *Mod. Phys. Lett. A* **2013**, *28*, 1350189. [CrossRef]
5. Tharanath, R.; Kuriakose, V.C. Thermodynamics and Spectroscopy of Schwarzschild black hole surrounded by Quintessence. *Mod. Phys. Lett. A* **2013**, *28*, 1350003. [CrossRef]
6. Carleo, A.; Lambiase, G.; Mastrototaro, L. Energy extraction via magnetic reconnection in Lorentz breaking Kerr–Sen and Kiselev black holes. *Eur. Phys. J. C* **2022**, *82*, 776. [CrossRef]
7. Toshmatov, B.; Stuchlík, Z.; Ahmedov, B. Rotating black hole solutions with quintessential energy. *Eur. Phys. J. Plus* **2017**,*132*, 98. [CrossRef]
8. Xu, Z.; Liao, Y.; Wang, J. Thermodynamics and phase transition in rotational Kiselev black hole. *arXiv* **2016**, arXiv:1610.00376. [CrossRef]
9. Kraus, P.; Wilczek, F. Self-interaction correction to black hole radiance. *Nucl. Phys. B* **1995**, *433*, 403. [CrossRef]
10. Visser, M. The Kiselev black hole is neither perfect fluid, nor is it quintessence. *Class. Quantum Gravity* **2020**, *37*, 045001. [CrossRef]
11. Parikh, M.K.; Wilczek, F. Hawking Radiation As Tunneling. *Phys. Rev. Lett.* **2000**, *85*, 5042. [CrossRef]
12. Parikh, M.K. *The Tenth Marcal Grossmann Meeting*; World Scientific: Singapore, 2006; pp. 1585–1590.
13. Zhang, J.Y.; Zhao, Z. Charged particles' tunnelling from the Kerr–Newman black hole. *Phys. Lett. B* **2006**, *638*, 110–113. [CrossRef]
14. Kerner, R.; Mann, R.B. Fermions tunnelling from black holes. *Class. Quantum Gravity***2008**,*25*, 095014. [CrossRef]
15. Jiang, Q.Q.; Wu, S.Q.; Cai, X. Hawking radiation from dilatonic black holes via anomalies. *Phys. Rev. D* **2007**, *75*, 064029. [CrossRef]
16. Srinivasan, K.; Padmanabhan, T. Particle production and complex path analysis. *Phys. Rev. D* **1999**, *60*, 024007. [CrossRef]
17. Shankaranarayanan, S.; Padmanablan, T.; Srinivasan, K. Hawking radiation in different coordinate settings: Complex paths approach. *Class. Quantum Gravity* **2002**, *19*, 2671. [CrossRef]
18. Kerner, R.; Mann, R.B. Charged fermions tunnelling from Kerr–Newman black holes. *Phys. Lett. B* **2008**, *665*, 277. [CrossRef]
19. Lin, K.; Yang, S.Z. Fermion tunneling from higher-dimensional black holes. *Phys. Rev. D* **2009**, *79*, 064035. [CrossRef]
20. Lin, K.; Yang, S.Z. Fermions tunneling of higher-dimensional Kerr–Anti-de Sitter black hole with one rotational parameter. *Phys. Lett. B* **2009**, *674*, 127. [CrossRef]
21. Yang, S.Z.; Lin, K. Hamilton-Jacobi Equation and Tunneling Radiation from Kerr-TAUB-NUT Black Holes. *Sci. Sin-Phys. Mech. As.* **2010**, *40*, 507.
22. Amelino-Camelia, G. Improved limit on quantum-spacetime modifications of Lorentz symmetry from observations of gamma-ray blazars. *New J. Phys.* **2004**, *6*, 188. [CrossRef]
23. Magueijo, J.; Smolin, L. Generalized Lorentz invariance with an invariant energy scale. *Phys. Rev. D* **2003**, *67*, 044017. [CrossRef]

24. Ellis, J.; Mavromatos, N.E.; Nanopoulos, D.V. A microscopic Liouville arrow of time. *Chao Solitons Fractals* **1999**, *10*, 345. [CrossRef]
25. Kruglov, S.I. Modified Dirac equation with Lorentz invariance violation and its solutions for particles in an external magnetic field. *Phys. Lett. B* **2012**, *718*, 228. [CrossRef]
26. Gomes, M.; Nascimento, J. R.; Petrov, A. Y.; da Silva, A. J. Aetherlike Lorentz-breaking actions. *Phys. Rev. D* **2010**, *81*, 045018. [CrossRef]
27. Nascimento, J. R.; Petrov, A. Y.; Reyes, C.M. Lorentz-breaking theory with higher derivatives in spinor sector. *Phys. Rev. D* **2015**, *92*, 045030. [CrossRef]
28. Yang, S.Z.; Lin, K.; Li, J.; Jiang, Q.Q. Lorentz Invariance Violation and Modified Hawking Fermions Tunneling Radiation. *Adv. High Energy Phys.* **2016**, *2016*, 7058764. [CrossRef]
29. Yang, S.Z.; Lin, K. Hawking tunneling radiation in Lorentz-violating scalar field theory. *Acta Phys. Sin.* **2019**, *68*, 060401. [CrossRef]
30. Wang, C.; Zhang, J.; Liu, Y.-Z. Modification Study on Quantum Tunneling Radiation of Kinnersley Black Hole. *Universe* **2023**, *9*, 496. [CrossRef]
31. Wang, C.; Zhang, J.; Yang, S.Z. Entropy Modifications of Charged Accelerating Anti-de Sitter Black Hole. *Entropy* **2025**, *27*, 900. [CrossRef] [PubMed]
32. Sha, B.; Liu, Z.-E. Lorentz-breaking theory and tunneling radiation correction to Vaidya–Bonner de Sitter Black Hole. *Eur. Phys. J. C* **2022**, *82*, 648. [CrossRef]
33. Li, R.; Yu, Z.H.; Yang, S.Z. Modification of Bekenstein-Hawking entropy of cylindrical symmetric black hole by Lorentz breaking theory. *EPL* **2023**, *141*, 50001. [CrossRef]
34. Li, R.; Ding, Q.T.; Yang, S.Z. Modified Hawking temperature and entropy of general stationary black holes by Lorentz invariance violation. *EPL* **2022**, *138*, 60001. [CrossRef]
35. Banerjee, R.; Majhi, B.R. Quantum tunneling beyond semiclassical approximation. *JHEP* **2008**, *6*, 95. [CrossRef]
36. Banerjee, R.; Majhi, B.R. Quantum tunneling and back reaction. *Phys. Lett. B* **2008**, *662*, 62–65. [CrossRef]
37. Feng, Z.W.; Li, H.L.; Zu, X.T.; Yang, S.Z. Corrections to the thermodynamics of Schwarzschild-Tangherlini black hole and the generalized uncertainty principle. *arXiv* **2015**, arXiv:1512.09219. [CrossRef]
38. Melia, F. The cosmic equation of state. *Astrophys. Space Sci.* **2015**, *356*, 393–398. [CrossRef]
39. Rizwan, M.; Jamil, M.; Wang, A. Distinguishing a rotating Kiselev black hole from a naked singularity using the spin precession of a test gyroscope. *Phys. Rev. D* **2018**, *98*, 024015. [CrossRef]

entropy MDPI

Article

Modified Gravity in the Presence of Matter Creation: Scenario for the Late Universe

Giovanni Montani [1,2,*], Nakia Carlevaro [1] and Mariaveronica De Angelis [3]

1 Nuclear Department, ENEA, C.R. Frascati, Via E. Fermi 45, 00044 Frascati, Italy; nakia.carlevaro@enea.it
2 Physics Department, "Sapienza" University of Rome, P. le A. Moro 5, 00185 Rome, Italy
3 School of Mathematics and Statistics, University of Sheffield, Hounsfield Road, Sheffield S3 7RH, UK; mdeangelis1@sheffield.ac.uk
* Correspondence: giovanni.montani@enea.it

Abstract: We consider a dynamic scenario for characterizing the late Universe evolution, aiming to mitigate the Hubble tension. Specifically, we consider a metric $f(R)$ gravity in the Jordan frame which is implemented to the dynamics of a flat isotropic Universe. This cosmological model incorporates a matter creation process, due to the time variation of the cosmological gravitational field. We model particle creation by representing the isotropic Universe (specifically, a given fiducial volume) as an open thermodynamic system. The resulting dynamical model involves four unknowns: the Hubble parameter, the non-minimally coupled scalar field, its potential, and the energy density of the matter component. We impose suitable conditions to derive a closed system for these functions of the redshift. In this model, the vacuum energy density of the present Universe is determined by the scalar field potential, in line with the modified gravity scenario. Hence, we construct a viable model, determining the form of the $f(R)$ theory a posteriori and appropriately constraining the phenomenological parameters of the matter creation process to eliminate tachyon modes. Finally, by analyzing the allowed parameter space, we demonstrate that the Planck evolution of the Hubble parameter can be reconciled with the late Universe dynamics, thus alleviating the Hubble tension.

Keywords: modified gravity; Late Universe; Hubble tension; classical cosmology

1. Introduction

Over the past decade, cosmological studies of the late Universe have uncovered a significant discrepancy in data, known as the "Hubble tension". This tension arises from the differing values of the Hubble constant (H_0) measured by the SH0ES Collaboration using Type Ia Supernova (SNIa) data [1–4] and those obtained by the Planck Satellite Collaboration [5]. The discrepancy, approximately 5σ, presents a perplexing challenge, prompting new considerations regarding the dynamics of the late Universe. For a series of discussions on this topic, see [6–15]. We also underline how, nonetheless, in the presence of such a tension, the most commonly accepted cosmological model which describes the evolution of the Hubble parameter is the so-called Lambda Cold Dark Matter (ΛCDM) model [16,17]. This corresponds to including, in the Universe dynamics, a matter-like contribution and a cosmological constant. The possibility of interpreting the Hubble tension as a continuous effective variation of the Hubble constant based on redshift—where its value appears to depend on the redshift region of the astrophysical sources used for its determination—has been supported by analyses in [18–20], and is also discussed in [21,22]. Moreover, the coexistence of SNIa and baryon acoustic oscillations (BAOs) [23] (and references therein) data within the same redshift region, and their resulting tension, as BAO provides a H_0 value compatible with that from Planck, has led to the development of early dark energy (DE) models [24] and a suitable combination of early and late modified dynamics [25,26]. In particular, the analysis in [7] proposed a specific $f(R)$ model, examined within the so-called Jordan frame [27,28], to effectively address the Hubble tension in the presence of evolutionary DE, which transits to a phantom contribution at $z > 1$. This model presents an intriguing scenario where the Hubble tension is essentially resolved at $z \gtrsim 2$, since the non-minimally coupled scalar field shows a monotonically increasing behaviour toward an asymptote. This asymptotic configuration aligns with the Planck value for the Hubble constant, occurring at a relatively low redshift.

In this work, we explore a similar approach, but rather than initially assuming the existence of evolutionary dark energy (DE), we consider a more natural physical scenario. This scenario involves the gravitational field generating weakly massive particles due to its time variation, effectively introducing a radiation component in the late Universe. Specifically, the matter creation process is treated phenomenologically, by regarding the Universe as an open thermodynamic system, as discussed in previous works like [29,30]. The particle creation rate is determined using an *ansatz* in the form of a power-law in the Hubble parameter. For a more comprehensive understanding of the matter creation process across the Universe, please see [31–34].

After constructing the dynamical system governing the late Universe dynamics and incorporating the creation of a radiation component by the gravitational background, we demonstrate how the Hubble tension could be effectively alleviated within certain favourable regions of the parameter space. This framework offers a promising depiction of the underlying mechanism potentially resolving such a cosmological issue. It is noteworthy that the current-day radiation created does not exceed a few percent, and the anticipated weakly interacting nature of these particles explains why this generated energy density remains indirectly observed in the actual Universe.

The paper is structured as follows. In Section 2, a phenomenological approach to the matter creation process is presented and the dynamics for the created energy density is fixed. In Section 3, the formulation of a metric $f(R)$ gravity in the Jordan frame is reviewed and the basic features of the proposed model are traced. In Section 4, the dynamical model corresponding to the evolution of a flat isotropic Universe in the considered modified scenario is constructed. The set of free parameters is described and the initial conditions for constructing the numerical solutions are given. In Section 5, the free parameter space is numerically investigated generating a triangular plot. A privileged set of parameters is then identified and the profile of the Hubble parameter is provided in order to show the capability of the model to alleviate the Hubble tension.

2. Matter Creation Process

In the thermodynamic framework presented in this work, the concept of matter creation from a time-varying gravitational field relies on a simple phenomenological representation. Let us start from the first principle, as follows:

$$dU = -pdV + \delta Q + \mu dN\,, \tag{1}$$

which has to be combined with the heat change δQ, expressed by means of the second thermodynamics principle as $\delta Q = TdS$. Here, U is the internal energy, p is the pressure, T is the temperature, S is the entropy, μ is the chemical potential, N is the particle number, and V is the volume of the considered system. Introducing now the expressions $U = \rho V$, $S = \sigma N$, and $\mu = (\rho + p)V/N - \sigma T$ (in which ρ is the total energy density of the system and σ is the entropy per particle), we can rewrite Equation (1) as follows:

$$d\rho = -(\rho + p)\left(1 - \frac{d\ln N}{d\ln V}\right)\frac{dV}{V} + T\frac{N}{V}d\sigma\,. \tag{2}$$

The difference from an iso-entropic system is that we only need to ensure the preservation of entropy per particle by setting $d\sigma = 0$. Hence, it is clear that, since $\sigma = S/N$, matter creation results in an increase in the entropy of the system. Therefore, Equation (2) simplifies as follows:

$$\frac{d\rho}{d\tau} = -(\rho + p)\left(1 - \frac{d\ln N}{d\ln V}\right)\frac{d\ln V}{d\tau}\,, \tag{3}$$

once it is applied to a homogeneous background and its evolution is tracked with respect to a clock labelled τ. Here, the time variable τ thus denotes the specific time variable associated with the considered dynamical system and, in what follows, it will be identified with the cosmological Gaussian time. The line element of a flat isotropic Universe [35] reads as follows:

$$ds^2 = -dt^2 + a^2(t)\left(dx^2 + dy^2 + dz^2\right), \quad (4)$$

in which t denotes the synchronous time ($c = 1$), and $\{x, y, z\}$ are the Cartesian coordinates. Moreover, $a(t)$ stands for the cosmic scale factor responsible for the Universe expansion.

In terms of the Hubble parameter $H(t) \equiv \dot{a}/a$ (the dot denoting synchronous time differentiation), a reliable *ansatz* for the matter creation is as follows:

$$\frac{d \ln N}{d \ln V} = (H/\bar{H})^{\beta}, \quad (5)$$

where β and $\bar{H}$ are positive free parameters of the model and the considered fiducial volume can be set as $V = a(t)^3 V_0$, with V_0 as a generic coordinate volume which does not enter the dynamics.

Since the present-day Universe expansion rate is rather slow, it is natural to argue that the gravitational field is generating very low-mass particles [36,37]. Given that, we address this process to a radiation-like component energy density ρ_r. According to the analysis above, the dynamics of such an emerging radiation contribution is governed by the following equation (i.e., we use the equation of state $p = w\rho$ with $w = 1/3$):

$$\dot{\rho}_r = -4H\rho_r\left(1 - (H/\bar{H})^{\beta}\right). \quad (6)$$

This equation comes from Equation (3), once we use $\tau = t$, the *ansatz* in Equation (5), and by implementing $d \ln V/dt = 3H$. In the following sections, we embed this mechanism in the context of a metric $f(R)$ gravity in the Jordan frame.

3. Modified Cosmological Dynamics in the Jordan Frame

In the Jordan frame, the action of a metric $f(R)$ gravity can be cast as follows [27]:

$$\mathcal{S} = \frac{1}{2\chi} \int d^4x \sqrt{-g}(\phi R - V(\phi)), \quad (7)$$

where χ denotes the Einstein constant, g and R are the metric determinant and the Ricci scalar, respectively, while the non-minimally coupled scalar field ϕ is the independent degree of freedom expressing the modified gravity formulation. In particular, the potential term $V(\phi)$ is linked to the specific form of the function $f(R)$ via the following relation:

$$f(R(\phi)) = \phi \frac{dV}{d\phi} - V(\phi), \quad (8)$$

which comes from the basic definition $\phi \equiv df/dR$ and from the substitution of the field equation into the expression of V, obtained by varying the action Equation (7) with respect to ϕ, i.e., $R = dV/d\phi$. A basic viability condition for the choice of a given $f(R)$ model is that the scalar mode possesses a positively defined quadratic mass, according to the following definition [38]:

$$\mu_{\phi}^2 \equiv \frac{1}{3}\left(\phi \frac{d^2V}{d\phi^2} - \frac{dV}{d\phi}\right) \geq 0. \quad (9)$$

The variation in the action (7) with respect to the metric tensor yields the vacuum Einstein equations of the extended scalar–tensor theory. By taking the trace of these equations and incorporating the condition $R = dV/d\phi$, we can derive a Klein–Gordon-like equation for the scalar field. Introducing a matter source is straightforward; it involves a conserved energy–momentum tensor that represents the specific physical system under consideration.

The trace of this tensor also contributes to the Klein–Gordon-like equation for the scalar field, as discussed by [27].

We now specify the field equation for the line element of a flat isotropic Universe (see Equation (4)) in the presence of a co-moving total energy density ρ_{tot}. In particular, we consider the 00-component of the Einstein Equation [16], as follows:

$$H^2 = \frac{\chi\rho_{tot}}{3\phi} - H\frac{\dot{\phi}}{\phi} + \frac{V(\phi)}{6\phi}. \tag{10}$$

Here, $\rho_{tot} = \rho_m + \rho_r$, where ρ_m denotes the (dark and baryonic) matter component of the Universe, verifying the standard conservation law i.e., the divergenceless nature of the corresponding energy–momentum tensor, as follows:

$$\dot{\rho}_m + 3H\rho_m = 0 \rightarrow \rho_m = \rho_m^0(1+z)^3, \tag{11}$$

in which ρ_m^0 denotes the present-day value of ρ_m. We introduced the redshift variable $z(t) = 1/a(t) - 1$ (we set the present value of the cosmic scale factor equal to unity). It is worth noting that the evolution of the Universe matter component is not affected by the matter creation process. Actually, the radiation creation, as described in Equation (6), does not follow the divergenceless character of a prefect fluid with $w = 1/3$. However, it corresponds to a perfect fluid with the following time-varying equation of a state parameter:

$$w_r(t) = \frac{1}{3} - \frac{4}{3}\left(\frac{H}{\bar{H}}\right)^{\beta}. \tag{12}$$

It is the conservation law of a such an energy–momentum tensor which provides Equation (6). The second basic equation of the cosmological dynamics reads as follows:

$$\frac{dV}{d\phi} = R = 12H^2 + 6\dot{H}. \tag{13}$$

This approach is particularly suitable for determining a posteriori the form of the $f(R)$ gravity that can help alleviate the Hubble tension.

4. Reduced Dynamics

In order to transform the potential $V(\phi)$ from an ingredient which is assigned via the function $f(R)$ into a dynamical variable $V(t)$, we impose on Equation (10) the following two conditions:

$$V(\phi) = 2\chi\rho_\Lambda + G(\phi), \qquad 6H\dot{\phi} = G, \tag{14}$$

where ρ_Λ is the constant value of the Universe energy density and G is a generic functional form, to be dynamically determined. Clearly, these two relations play a crucial role in giving Equation (10) a form that resembles the dynamics of a ΛCDM model for the Universe, as specified at the beginning of the next section. Introducing the critical parameters in place of the energy density according to the relations $\Omega_m^0 = \chi\rho_m/3H_0^2$, $\Omega_\Lambda^0 = \chi\rho_\Lambda/3H_0^2$, and $\Omega_r = \chi\rho_r/3H_0^2$, where H_0 is the present day value of the Hubble constant, we can rewrite Equation (10) into the following form:

$$H^2(z) = \frac{H_0^2}{\phi(z)}\left(\Omega_m^0(1+z)^3 + \Omega_\Lambda^0 + \Omega_r(z)\right), \tag{15}$$

where $\Omega_r^0 = \Omega_r(z = 0)$ and the relation $\Omega_\Lambda^0 = 1 - \Omega_m^0 - \Omega_r^0$ must be satisfied, given that we set $\phi(z = 0) = 1$. This equation is coupled with the dynamical system of Equations (6), (13), and (14) recast as follows (the prime denotes z-differentiation):

$$\phi' = -\frac{G(z)}{6(1+z)H^2}, \tag{16}$$

$$G' = 3\phi'\left(4H^2 - 2(1+z)HH'\right), \tag{17}$$

$$\Omega_r' = 4(1+z)\Omega_r\left(1 - (H/\bar{H})^\beta\right). \tag{18}$$

The ratio of Equations (16) and (17) gives us the following relation [7]:

$$G(z) = -6A^2 H(z)(1+z)^{-2}, \tag{19}$$

where A^2 is a positive integration constant. Substituting the expression above into Equation (16), we obtain the following:

$$\phi' = \frac{A^2}{H(z)(1+z)^3}, \tag{20}$$

which describes the increasing behaviour of ϕ with the time variable z.

In this scheme, the dynamical system thus reduces to the following:

$$\phi' = A_0^2\,(1+z)^{-3}\,\left(\phi(z)^{-1}(\Omega_m^0(1+z)^3 + 1 - \Omega_m^0 - \Omega_r^0 + \Omega_r(z))\right)^{-1/2}, \tag{21}$$

$$\Omega_r' = 4(1+z)^{-1}\Omega_r(x)\left(1 - (\bar{H}_0^{-2}\phi(z)^{-1}(\Omega_m^0(1+z)^3 + 1 - \Omega_m^0 - \Omega_r^0 + \Omega_r(z)))^{\beta/2}\right), \tag{22}$$

where $A_0^2 \equiv A^2/H_0$ and $\bar{H}_0 \equiv \bar{H}/H_0$. Once we have calculated the function $\phi(z)$, we can also determine the potential $G(\phi)$ and, finally, the form of the resulting $f(R)$ gravity.

5. Numerical Analysis

In this section, we provide a comprehensive description of the methodology used for integrating the model. Specifically, we numerically derive the dynamical forms of $\phi(z)$ and $\Omega_r(z)$, which ultimately lead to the final expression of $H(z)$ in Equation (15). To compare our results with the standard two flat ΛCDM forms of the Hubble parameter constrained by the early Universe data [5] and by the Phanteon+ dataset [3], we introduce the following quantities:

$$H_{Pl}(z) = H_0^{Pl}(\Omega_m^{0Pl}(1+z)^3 + (1 - \Omega_m^{0Pl}))^{1/2}, \tag{23}$$

$$H_{SN}(z) = H_0^{SN}(\Omega_m^{0SN}(1+z)^3 + (1 - \Omega_m^{0SN}))^{1/2}, \tag{24}$$

with the following (H and H_0 are in units of km s^{-1} Mpc^{-1}):

$$H_0^{Pl} = 67.4 \pm 0.5\,, \qquad \Omega_m^{0Pl} = 0.315 \pm 0.007\,, \tag{25}$$

$$H_0^{SN} = 73.6 \pm 1.1\,, \qquad \Omega_m^{0SN} = 0.334 \pm 0.018\,, \tag{26}$$

In order to study the viability of the addressed scenario and also its capability in alleviating the Hubble tension, we explore the full free parameter space of the model $\{H_0, \Omega_m^0, \Omega_r^0, \bar{H}_0, A_0, \beta\}$. Equations (21) and (22) are numerically solved, spanning a grid of 15 values for each parameter, and thus collecting 15^6 sampled different models. The ranges are defined according to previously conducted tests on the integrability of the system and phenomenological considerations: we have set $72.5 \le H_0 \le 74.7$, $0.25 \le \Omega_m^0 \le 0.335$, $0.01 \le \Omega_r^0 \le 0.15$, $0.5 \le \bar{H}_0 \le 1.5$, $0.1 \le A_0 \le 0.5$, and $0.5 \le \beta \le 1.5$. We note that the selected values of H_0 are in accordance with the measurements of the Pantheon+ analysis [3], while Ω_r^0 is assumed, as already stated, to remain a small contribution to the energy density of the Universe.

The obtained models are then filtered requiring non-tachyonic modes, i.e., that Equation (9) is guaranteed, and we further impose the condition that the normalized (by H_0^2) squared mass is less or equal to unity, today. This point is relevant in order for the considered $f(R)$ model to obey the so-called "chameleon" mechanism [39]. The resulting models are thus physically viable and, to study the efficiency in order to alleviate the tension, we also require that $0.999 < H/H_{Pl}|_{z=10} < 1.001$. With this procedure, we finally obtain around 5×10^3 sampled models. We then convolve the data with a normal distribution to create a smooth density estimate. A kernel density estimate plot is a visualization method used to depict the distribution of observations in a dataset, akin to a histogram. The results are depicted in Figure 1.

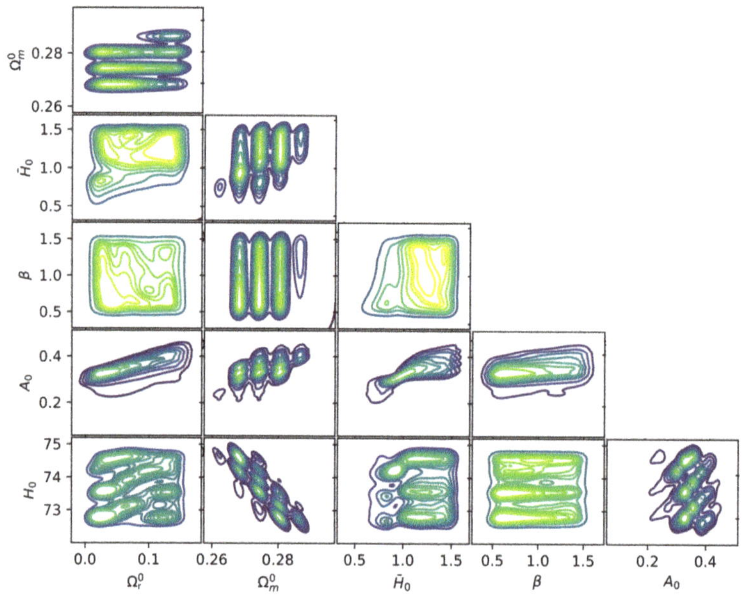

Figure 1. Density plot for all the model's parameters. The yellow region indicates the most frequent values preferred by the model.

From this analysis, preferred regions can be individualized providing the most frequent parameter sets. As an example of the model's capability in alleviating the Hubble tension, we select the following:

$$H_0 = 72.9\,, \quad \Omega_m^0 = 0.28\,, \quad \Omega_r^0 = 0.035\,, \quad \bar{H}_0 = 1.01\,, \quad A_0 = 0.33\,, \quad \beta = 0.7\,. \tag{27}$$

In Figure 2, we plot the resulting evolution of $H(z)$ from Equation (15), with this parameter setup, together with the curves in Equation (23) (blue) and Equation (24) (red), considering the corresponding errors. The tension alleviation emerges clearly, and the $H(z)$ profile overlaps $H_{Pl}(z)$ already at $z \simeq 3$, but reaches higher values of the Hubble constant in $z = 0$. In the figure, we also depict six relevant measurements for BAO sources (in the range $0.3 < z < 3$) [40–43] and the SH0ES prior for $H(z)$ today. This clearly indicates the capability of the addressed model to also alleviate the tension derived from different late Universe sources. In Figure 3, we instead plot the functions ϕ and Ω_r as functions of z. The curves are obtained by integrating Equations (21) and (22), assuming the choice of the parameters as in Equation (27). As expected, for $z > 3$, $\Omega_r(z)$ approaches zero, while $\phi(z)$ is frozen as a plateau.

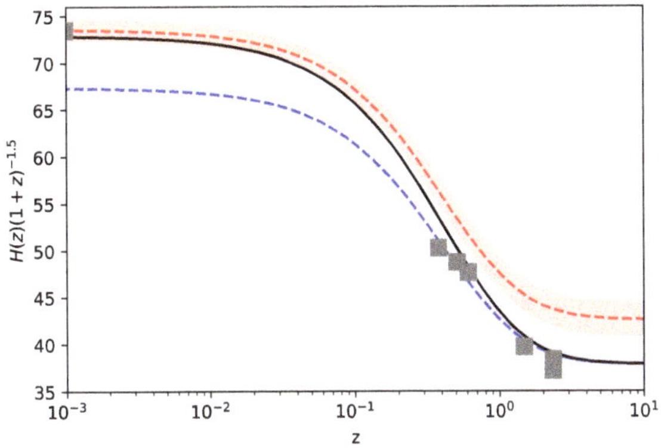

Figure 2. Plot of $H(z)$ (black) using the parameters in Equation (27) and the profiles (with the corresponding errors) of $H_{Pl}(z)$ (blue) and $H_{SN}(z)$ (red). Grey squares represent the SH0ES prior and six relevant measurements for BAO sources for $0.3 < z < 3$.

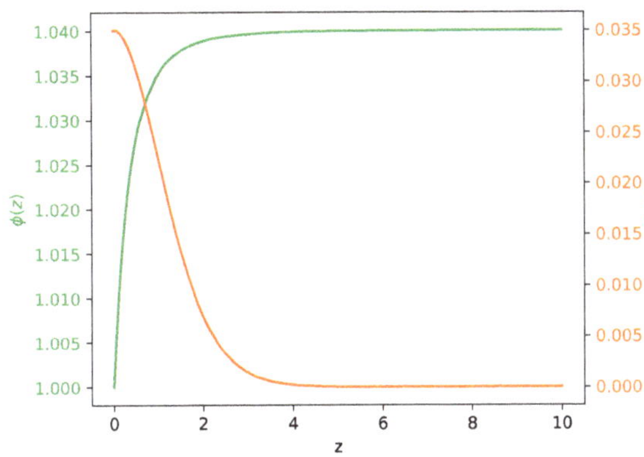

Figure 3. Plot of $\phi(z)$ (green) and $\Omega_r(z)$ (orange) integrated from Equations (21) and (22), implementing the parameters in Equation (27).

This study of the parameter space is essential for guiding the real data analysis procedure, clearly indicating the model's viability for interpreting the Hubble tension through the missing physics that must be added to the ΛCDM formulation to reconcile it with observations. The numerical study highlights how the additional components of modified metric $f(R)$ gravity and the radiation term generated by the cosmological background are crucial for mitigating the Hubble tension within the framework of a physical theory. In this context, a key aspect of the numerical filtering procedure is ensuring that the modified gravity is not associated with a tachyonic mode.

6. Conclusions

We constructed a revised dynamical model for the late Universe based on both a metric $f(R)$ gravity in the Jordan frame, similar to the approach in [7], and a phenomenological model for matter creation associated with the time variation of the cosmological gravitational field. The matter creation is described by treating the Universe as an open thermodynamic system, with the rate of particle creation phenomenologically regulated by a power-law of the Hubble parameter. The form of the $f(R)$ gravity is not assigned a priori. Indeed, by imposing suitable conditions on the generalized Friedmann equation, the potential of the non-minimally coupled scalar field $V(z) \equiv V(\phi(z))$ becomes a dynamical variable, along with $\phi(z)$ and $\Omega_r(z)$. Consequently, it is possible to reconstruct the form of $V(\phi)$ and thus determine the function $f(R)$ governing the modified gravity.

We developed a dynamical model with six free parameters, $\{H_0, \Omega_m^0, \Omega_r^0, \bar{H}_0, A_0, \beta\}$, aimed at alleviating the Hubble tension. Our numerical analysis involved a thorough screening of all possible solutions, retaining only those parameter sets that ensured the physical consistency of the proposed scenario. These solutions also had to meet two criteria: the resulting Hubble parameter needed to exhibit the required asymptotic ΛCDM behaviour (achieved at low z values) and allow for a Hubble constant compatible with the SH0ES collaboration observations. Identifying a preferred set of parameters provided significant insights into how real data comparisons could impact the parameter space. This approach could successfully address the Hubble tension, as the Hubble parameter rapidly converges to the ΛCDM model with Planck-measured parameters as the redshift increases.

We conclude by emphasizing that the privileged values of Ω_r^0 were found to be very small, making this component reliably unobserved through direct measurement. The particles forming this radiation component, such as sterile neutrinos [44], very weakly interact, which explains their elusive nature. The key takeaway from this study is that the Hubble tension can be effectively addressed by combining metric $f(R)$ gravity with additional modifications to the standard ΛCDM model.

Author Contributions: Conceptualization, G.M.; methodology, G.M., N.C. and M.D.A.; software, N.C. and M.D.A.; validation, N.C. and M.D.A.; formal analysis, G.M., N.C. and M.D.A.; investigation, G.M., N.C. and M.D.A.; resources, N.C. and M.D.A.; data curation, N.C. and M.D.A.; writing—original draft preparation, G.M., N.C. and M.D.A.; writing—review and editing, G.M., N.C. and M.D.A.; visualization, N.C. and M.D.A. All authors have read and agreed to the published version of the manuscript.

Funding: This research received no external funding.

Institutional Review Board Statement: Not applicable.

Data Availability Statement: The original contributions presented in the study are included in the article; further inquiries can be directed to the corresponding author.

Acknowledgments: M.D.A. thanks William Giarè for valuable discussions.

Conflicts of Interest: The authors declare no conflicts of interest.

References

1. Scolnic, D.M.; Brout, D.; Carr, A.; Riess, A.G.; Davis, T.M.; Dwomoh, A.; Jones, D.O.; Ali, N.; Charvu, P.; Chen, R.; et al. The Pantheon+ Analysis: The Full Data Set and Light-curve Release. *Astrophys. J.* **2022**, *938*, 113. [CrossRef]
2. Scolnic, D.M.; Jones, D.O.; Rest, A.; Pan, Y.C.; Chornock, R.; Foley, R.J.; Huber, M.E.; Kessler, R.; Narayan, G.; Riess, A.G.; et al. The Complete Light-curve Sample of Spectroscopically Confirmed SNe Ia from Pan-STARRS1 and Cosmological Constraints from the Combined Pantheon Sample. *Astrophys. J.* **2018**, *859*, 101. [CrossRef]
3. Brout, D.; Scolnic, D.; Popovic, B.; Riess, A.G.; Carr, A.; Zuntz, J.; Kessler, R.; Davis, T.M.; Hinton, S.; Jones, D.; et al. The Pantheon+ Analysis: Cosmological Constraints. *Astrophys. J.* **2022**, *938*, 110. [CrossRef]
4. Riess, A.G.; Yuan, W.; Macri, L.M.; Scolnic, D.; Brout, D.; Casertano, S.; Jones, D.O.; Murakami, Y.; Anand, G.S.; Breuval, L.; et al. A Comprehensive Measurement of the Local Value of the Hubble Constant with 1 km s^{-1} Mpc^{-1} Uncertainty from the Hubble Space Telescope and the SH0ES Team. *Astrophys. J. Lett.* **2022**, *934*, L7. [CrossRef]
5. Aghanim, N.; Akrami, Y.; Ashdown, M.; Aumont, J.; Baccigalupi, C.; Ballardini, M.; Banday, A.J.; Barreiro, R.B.; Bartolo, N.; Basak, S.; et al. Planck 2018 results. VI. Cosmological parameters. *Astron. Astrophys.* **2021**, *641*, A6. Erratum in *Astron. Astrophys.* **2021**, *652*, C4. [CrossRef]
6. Schiavone, T.; Montani, G.; Bombacigno, F. f(R) gravity in the Jordan frame as a paradigm for the Hubble tension. *Mon. Not. RAS* **2023**, *522*, L72–L77. [CrossRef]
7. Montani, G.; De Angelis, M.; Bombacigno, F.; Carlevaro, N. Metric f(R) gravity with dynamical dark energy as a scenario for the Hubble tension. *Mon. Not. RAS* **2024**, *527*, L156–L161. [CrossRef]
8. Montani, G.; Carlevaro, N.; Dainotti, M.G. Slow-rolling scalar dynamics as solution for the Hubble tension. *Phys. Dark Univ.* **2024**, *44*, 101486. [CrossRef]
9. Adil, S.; Akarsu, Ö.; Di Valentino, E.; Nunes, R.; Özülker, E.; Sen, A.; Specogna, E. Omnipotent dark energy: A phenomenological answer to the Hubble tension. *Phys. Rev. D* **2024**, *109*, 023527. [CrossRef]

10. Nojiri, S.; Odintsov, S.D.; Oikonomou, V.K. Integral F(R) gravity and saddle point condition as a remedy for the H0-tension. *Nucl. Phys. B* **2022**, *980*, 115850. [CrossRef]
11. Odintsov, S.D.; Sáez-Chillón Gómez, D.; Sharov, G.S. Analyzing the H_0 tension in $F(R)$ gravity models. *Nucl. Phys. B* **2021**, *966*, 115377. [CrossRef]
12. Wang, B.; Abdalla, E.; Atrio-Barandela, F.; Pavón, D. Dark matter and dark energy interactions: Theoretical challenges, cosmological implications and observational signatures. *Rep. Prog. Phys.* **2016**, *79*, 096901. [CrossRef]
13. Naidoo, K.; Jaber, M.; Hellwing, W.A.; Bilicki, M. Dark matter solution to the H_0 and S_8 tensions, and the integrated Sachs-Wolfe void anomaly. *Phys. Rev. D* **2024**, *109*, 083511. [CrossRef]
14. Dainotti, M.; Bargiacchi, G.; Bogdan, M.; Capozziello, S.; Nagataki, S. On the statistical assumption on the distance moduli of Supernovae Ia and its impact on the determination of cosmological parameters. *J. High Energy Astrophys.* **2024**, *41*, 30–41. [CrossRef]
15. Dainotti, M.G.; Bargiacchi, G.; Bogdan, M.; Lenart, A.L.; Iwasaki, K.; Capozziello, S.; Zhang, B.; Fraija, N. Reducing the Uncertainty on the Hubble Constant up to 35% with an Improved Statistical Analysis: Different Best-fit Likelihoods for Type Ia Supernovae, Baryon Acoustic Oscillations, Quasars, and Gamma-Ray Bursts. *Astrophys. J.* **2023**, *951*, 63. [CrossRef]
16. Weinberg, S. *Cosmology*; OUP Oxford: Oxford, UK, 2008.
17. Di Valentino, E.; Mena, O.; Pan, S.; Visinelli, L.; Yang, W.; Melchiorri, A.; Mota, D.F.; Riess, A.G.; Silk, J. In the realm of the Hubble tension—a review of solutions. *Class. Quant. Grav.* **2021**, *38*, 153001. [CrossRef]
18. Dainotti, M.G.; De Simone, B.; Schiavone, T.; Montani, G.; Rinaldi, E.; Lambiase, G. On the Hubble constant tension in the SNe Ia Pantheon sample. *Astrophys. J.* **2021**, *912*, 150. [CrossRef]
19. Dainotti, M.G.; De Simone, B.; Schiavone, T.; Montani, G.; Rinaldi, E.; Lambiase, G.; Bogdan, M.; Ugale, S. On the Evolution of the Hubble Constant with the SNe Ia Pantheon Sample and Baryon Acoustic Oscillations: A Feasibility Study for GRB-Cosmology in 2030. *Galaxies* **2022**, *10*, 24. [CrossRef]
20. Xu, B.; Xu, J.; Zhang, K.; Fu, X.; Huang, Q. Model-independent test of the running Hubble constant from the Type Ia supernovae and the Hubble parameter data. *Mon. Not. RAS* **2024**, *530*, 5091–5098. [CrossRef]
21. Krishnan, C.; Colgáin, E.O.; Sheikh-Jabbari, M.M.; Yang, T. Running Hubble Tension and a H0 Diagnostic. *Phys. Rev. D* **2021**, *103*, 103509. [CrossRef]
22. Kazantzidis, L.; Perivolaropoulos, L. Hints of a Local Matter Underdensity or Modified Gravity in the Low *z* Pantheon data. *Phys. Rev. D* **2020**, *102*, 023520. [CrossRef]
23. Efstathiou, G. To H_0 or not to H_0? *Mon. Not. RAS* **2021**, *505*, 3866–3872. [CrossRef]
24. Giarè, W. Inflation, the Hubble Tension and Early Dark Energy: An alternative overview. *Phys. Rev. D* **2024**, *109*, 123545. [CrossRef]
25. Vagnozzi, S. Seven Hints That Early-Time New Physics Alone Is Not Sufficient to Solve the Hubble Tension. *Universe* **2023**, *9*, 393. [CrossRef]
26. Vagnozzi, S. New physics in light of the H_0 tension: An alternative view. *Phys. Rev. D* **2020**, *102*, 023518. [CrossRef]
27. Sotiriou, T.P.; Faraoni, V. $f(R)$ theories of gravity. *Rev. Mod. Phys.* **2010**, *82*, 451–497. [CrossRef]
28. Nojiri, S.; Odintsov, S.D. Unified cosmic history in modified gravity: From F(R) theory to Lorentz non-invariant models. *Phys. Rep.* **2011**, *505*, 59–144. [CrossRef]
29. Montani, G. Influence of particle creation on flat and negative curved FLRW universes. *Class. Quantum Gravity* **2001**, *18*, 193–203. [CrossRef]
30. Calvão, M.O.; Lima, J.A.S.; Waga, I. On the thermodynamics of matter creation in cosmology. *Phys. Lett. A* **1992**, *162*, 223–226. [CrossRef]
31. Erdem, R. Gravitational particle production and the Hubble tension. *arXiv* **2024**, arxiv:2402.16791.
32. Nunes, R.C.; Pavón, D. Phantom behavior via cosmological creation of particles. *Phys. Rev. D* **2015**, *91*, 063526. [CrossRef]
33. Nunes, R.C. Gravitationally induced particle production and its impact on structure formation. *Gen. Relativ. Gravit.* **2016**, *48*, 107. [CrossRef]
34. Lima, J.A.S.; Graef, L.L.; Pavón, D.; Basilakos, S. Cosmic acceleration without dark energy: Background tests and thermodynamic analysis. *J. Cosmol. Astropart. Phys.* **2014**, *2014*, 42. [CrossRef]
35. Efstathiou, G.; Gratton, S. The evidence for a spatially flat Universe. *Mon. Not. RAS* **2020**, *496*, L91–L95. [CrossRef]
36. Birrell, N.; Davies, P. *Quantum Fields in Curved Space*; Cambridge University Press: Cambridge, MA, USA, 1982.
37. Wald, R. *Quantum Field Theory in Curved Spacetime and Black Hole Thermodynamics*; The University of Chicago Press: Chicago, IL, USA, 1994.
38. Olmo, G.J. Post–Newtonian constraints on f(R) cosmologies in metric and Palatini formalism. *Phys. Rev. D* **2005**, *72*, 083505. [CrossRef]
39. Brax, P.; van de Bruck, C.; Davis, A.C.; Shaw, D.J. f(R) Gravity and Chameleon Theories. *Phys. Rev. D* **2008**, *78*, 104021. [CrossRef]
40. Alam, S.; Ata, M.; Bailey, S.; Beutler, F.; Bizyaev, D.; Blazek, J.A.; Bolton, A.S.; Brownstein, J.R.; Burden, A.; Chuang, C.H.; et al. The clustering of galaxies in the completed SDSS-III Baryon Oscillation Spectroscopic Survey: Cosmological analysis of the DR12 galaxy sample. *Mon. Not. RAS* **2017**, *470*, 2617–2652. [CrossRef]

41. Blomqvist, M.; du Mas des Bourboux, H.; Busca, N.G.; de Sainte Agathe, V.; Rich, J.; Balland, C.; Bautista, J.E.; Dawson, K. Font-Ribera, A.; Guy, J.; et al. Baryon acoustic oscillations from the cross-correlation of Lyα absorption and quasars in eBOSS DR14. *Astron. Astrophys.* **2019**, *629*, A86. [CrossRef]
42. de Sainte Agathe, V.; Balland, C.; du Mas des Bourboux, H.; Busca, N.G.; Blomqvist, M.; Guy, J.; Rich, J.; Font-Ribera, A.; Pieri M.M.; Bautista, J.E.; et al. Baryon acoustic oscillations at z = 2.34 from the correlations of Lyα absorption in eBOSS DR14. *Astron Astrophys.* **2019**, *629*, A85. [CrossRef]
43. Hou, J.; Sánchez, A.G.; Ross, A.J.; Smith, A.; Neveux, R.; Bautista, J.; Burtin, E.; Zhao, C.; Scoccimarro, R.; Dawson, K.S.; et al. The completed SDSS-IV extended Baryon Oscillation Spectroscopic Survey: BAO and RSD measurements from anisotropic clustering analysis of the quasar sample in configuration space between redshift 0.8 and 2.2. *Mon. Not. RAS* **2021**, *500*, 1201–1221. [CrossRef]
44. Abazajian, K.N. Sterile neutrinos in cosmology. *Phys. Rept.* **2017**, *711*, 1–28. [CrossRef]

entropy

MDPI

Article

Two Dynamical Scenarios for Binned Master Sample Interpretation

Giovanni Montani [1,2,*], Elisa Fazzari [2,3,4], Nakia Carlevaro [1] and Maria Giovanna Dainotti [5,6,7]

1 Nuclear Department, ENEA, C.R. Frascati, Via E. Fermi 45, 00044 Frascati, Italy; nakia.carlevaro@enea.it
2 Physics Department, "Sapienza" University of Rome, P.le A. Moro 5, 00185 Roma, Italy; elisa.fazzari@uniroma1.it
3 Istituto Nazionale di Fisica Nucleare (INFN), Sezione di Roma, P.le A. Moro 5, 00185 Roma, Italy
4 Physics Department, Tor Vergata University of Rome, Via della Ricerca Scientifica 1, 00133 Roma, Italy
5 Division of Science, National Astronomical Observatory of Japan, 2-21-1, Mitaka 181-8588, Tokyo, Japan; maria.dainotti@nao.ac.jp
6 Astronomy Department, The Graduate University for Advanced Studies (SOKENDAI), Shonankokusaimura, Hayama, Miura 240-0115, Kanagawa, Japan
7 Space Science Institutes, 4765 Walnut St Ste B, Boulder, CO 80301, USA
* Correspondence: giovanni.montani@enea.it

Abstract: We analyze two different scenarios for the late universe dynamics, resulting in Hubble parameters deviating from the ΛCDM, mainly for the presence of an additional free parameter, which is the dark energy parameter. The first model consists of a pure evolutionary dark energy paradigm as a result of its creation by the gravitational field of the expanding universe. The second model also considers an interaction of the evolutionary dark energy with the matter component, postulated via the conservation of the sum of their ideal energy–momentum tensors. These two models are then compared via the diagnostic tool of the effective running Hubble constant, with the binned data of the so-called "Master sample" for the Type Ia Supernovae. The comparison procedures, based on a standard MCMC analysis, lead to a clear preference of data for the dark energy–matter interaction model, which is associated with a phantom matter equation of state parameter (very close to -1) when, being left free by data (it has a flat posterior), it is fixed in order to reproduce the decreasing power-law behavior of the effective running Hubble constant, already discussed in the literature.

Keywords: dark energy; late universe; supernovae

1. Introduction

The emergence of a 4σ discrepancy between the measurement of the Hubble constant [1,2] by the SH0ES and Planck collaborations, refs. [3,4] referred to as Hubble tension, led the community to reconsider the ΛCDM model as the only viable cosmological dynamics for the late Universe evolution [3,5–37]. This tension is further emphasized by several other independent and model-independent determinations of H_0, distinct from SNe Ia calibrated with Cepheids (see [38] for a review). These include measurements based on SNe Ia calibrated with the tip of the red giant branch (TRGB) [39–41], galaxy distances using the Tully–Fisher relation (TFR) [42,43], Active Galactic Nuclei (AGN) [44], Gravitational Waves (GWs) and Dark Sirens [45–49], strong lensing time delays [50,51], Type II Supernovae [52], Megamasers [53], and Surface Brightness Fluctuations [54].

A fundamental step regarding the reliability of the ΛCDM model has been achieved by the DESI collaboration [55,56], which demonstrated how the redshift profile of their

Baryonic Acoustic Oscillation (BAO) data is better fitted by a Chavellier–Polarski–Linder (CPL) model [57,58], also dubbed the w_0w_aCDM scenario. This result de facto stated convincing evidence for an evolutionary dark energy component across the late universe (see also [59–66]). However, the DESI collaboration, while partially solving the tension between the BAO and Cosmic Microwave Background (CMB) data, left entirely open the question of calibration based on the acoustic sound horizon and that of SH0ES, using Cepheids as standard candles [3]. The resulting picture, emerging from the last ten years of cosmological studies, is rather confused, and it suggests that new physical effects must be included in the late universe dynamics in order for it to take a consistent shape when combining different sources belonging to different redshift regions.

In [67,68], see also [69,70], these considerations led to investigating whether, in the same Type Ia Supernovae (SNe Ia) redshift distribution, an effective dependence of the Hubble constant on the considered binned representation can arise. In fact, a phenomenological power-law decreasing behavior has been detected as a better fit to binned data with respect to the ΛCDM model. The power-law behavior seen in the H_0 could be due either to the dynamical evolving dark energy model or to the $f(R)$ theory of gravity (see [12] and also [71]), but it could also account for selection biases, which are hidden due to different statistical assumptions (see [72]) and could be due to the evolution of the parameters of the SNe Ia (see [73]), the tip of the red giant brach stars [74], and of GRBs [75].

2. Theoretical Furmulation

Here, we perform a study based on two different models, which are then compared with the binned data of the so-called "Master sample" [69], in analogy to the study in [70]. The first scenario is based on an evolutionary dark energy formulation, which evolves with the redshift because it is created by the background gravitational field of the expanding universe [15,70,76–78] (for a detailed discussion see [Schiavone et al., "Revisiting the Matter-Creation Process: Constraints from Late-Time Acceleration and the Hubble Tension", in preparation]), according to a phenomenological ansatz for the produced particle rate, slightly modified with respect to the studies in [79,80] (actually we propose a mixed version of the two rates discussed in these papers, respectively). The second model is a revised version of the previous one, but with a completely different physical picture. In fact, the dark energy is still created by the gravitational field, but now it also interacts with the universe matter component via the condition that the sum of their energy–momentum tensors is conserved, instead of treating them separately (for other approaches with deformed matter contribution with respect to the ΛCDM model, see [81]). After testing these two models with the binned Master sample, we compare their predictions, especially in comparison to the ΛCDM dynamics and the power-law scaling, respectively.

The general scenario for developing the late universe models is a flat isotropic picture [82,83], whose line element reads as

$$ds^2 = -dt^2 + a^2(t)dl^2 \,, \tag{1}$$

where t denotes the synchronous time, dl^2 is the Euclidean infinitesimal distance, and $a(t)$ is the cosmic scale factor, regulating the expansion of the universe. The cosmological dynamics is driven by a (cold dark ρ_{dm} and baryonic ρ_b) matter energy density $\rho_m = \rho_{dm} + \rho_b$ and a dark energy contribution, with energy density ρ_{de} (we neglect here the radiation energy density, to be restored when investigating the dynamics up to the recombination era). Hence, the Friedmann equation is stated as follows:

$$H^2(t) \equiv \left(\frac{\dot{a}}{a}\right)^2 = \frac{\chi}{3}(\rho_m(t) + \rho_{de}(t)) \,, \tag{2}$$

where the dot refers to differentiation with respect to t and χ denotes the Einstein constant.

In the first proposed model, we retain the matter contribution ρ_m in its standard form, i.e., governed by the dynamics $\dot{\rho}_m(t) + 3H\rho_m(t) = 0$, which provides

$$\rho_m(z) = \rho_{m0}(1+z)^3 \,, \tag{3}$$

where, here and in the whole paper, we denote with the subscript 0 the present-day value of a quantity, and we have introduced the redshift variable $z \equiv 1/a - 1$ (we set equal to unity the present-day scale factor value). As a modification with respect to the standard ΛCDM model, we consider a process of dark energy creation by the (time-varying) gravitational field of the expanding universe, which is assumed to take place at equilibrium. Under these hypotheses, the continuity equation for the dark energy density reads as follows:

$$\dot{\rho}_{de}(t) = -3H(t)(1+w_{de})\left(1 - \frac{\Gamma(H,\rho_{de})}{3H(t)}\right)\rho_{de}(t) \,, \tag{4}$$

where Γ is the particle creation rate [79,80,84]. Above, w_{de} is the dark energy parameter and it is a free parameter of the model, subject to the constraint $w_{de} < -1/3$. In general, the phenomenological function $\Gamma(H,\rho_{de})$ is taken as a power-law of its own arguments [76,79]. In this respect, here, we consider the following *ansatz*:

$$\Gamma(H,\rho_{de}) = \Gamma^* H \rho_{de}^{-\alpha} \,, \tag{5}$$

where Γ^* and α are positive constants. We consider in the rate expression a linear term in H, stating, at highest order, the role of time-varying gravity in creating particles. We also add a dependence of Γ on a negative power of the dark energy density to suppress particle creation when its energy density increases.

Introducing the normalization $\Omega_{de} \equiv \chi\rho_{de}/3H_0^2$, where H_0 denotes the Hubble constant, Equation (4) can be rewritten as

$$\Omega'_{de}(z) = 3(1+w_{de})(1 - \bar{\Gamma}\,\Omega_{de}^{-\alpha})\Omega_{de}/(1+z) \,, \tag{6}$$

where the prime indicates differentiation with respect to the redshift z and we have defined $\bar{\Gamma} = (\Gamma^*/3)(\chi/3H_0^2)^\alpha$. Furthermore, accordingly defining $\Omega_{m0} = \chi\rho_{m0}/3H_0^2$, the Friedmann equation Equation (2), via Equation (3), stands as follows:

$$E^2(z) \equiv \left(\frac{H}{H_0}\right)^2 = \Omega_{m0}(1+z)^3 + \Omega_{de}(z) \,, \tag{7}$$

with $E(z)$ denoting the universe expansion rate. Thus we get the following initial condition $\Omega_{de}(0) = 1 - \Omega_{m0}$, and Equation (6) now admits the following solution:

$$\Omega_{de}(z) = \left[\bar{\Gamma} + ((1-\Omega_{m0})^\alpha - \bar{\Gamma})\,(1+z)^{3(1+w_{de})\alpha}\right]^{1/\alpha} . \tag{8}$$

We finally get, associated with our first model, the following Hubble parameter:

$$H(z) = H_0\sqrt{\Omega_{m0}(1+z)^3 + \left[\bar{\Gamma} + ((1-\Omega_{m0})^\alpha - \bar{\Gamma})(1+z)^{3(1+w_{de})\alpha}\right]^{1/\alpha}} . \tag{9}$$

This evolutionary dark energy model generalizes the ΛCDM dynamics, and it contains five free parameters, i.e., H_0, Ω_{m0}, $\bar{\Gamma}$, α, and w_{de}. It is immediately recognizable that the Hubble parameter in Equation (9) reduces to the standard ΛCDM one when we require $w_{de} = -1$. Finally, it is easy to realize that for $z > 1$ the $H(z)$ is very weakly sensitive to the values taken by the parameter α, and, in what follows, we will address the simplest case $\alpha = 1$.

It is worth noting that, in this case, the effective equation of state for the dark energy is associated, in $z = 0$, to the following parameter w_0^{eff}:

$$w_0^{eff} = -1 + (1 + w_{de})\left(1 - \frac{\bar{\Gamma}}{\Omega_{de}(0)}\right). \tag{10}$$

Now we require that, today, the intrinsic nature of dark energy, traced by w_{de}, remains the same, quintessence or phantom, also for the effective equation of state, dictated by w_0^{eff}. To this end, we have to require the constraint $\bar{\Gamma}\Omega_{de}(0) < 1$. Since the value of $\Omega_{de}(0)$ is expected (see also the concordance hypothesis [4]) close to the value 0.7, in what follows, we will consider for $\bar{\Gamma}$ the reference value 0.5.

Let us now introduce the diagnostic tool, corresponding to the effective running Hubble constant with the redshift $\mathcal{H}_0(z)$ (see [67–71]), i.e.,

$$\mathcal{H}_0(z) \equiv \frac{H(z)}{\sqrt{\Omega_m^0(1+z)^3 + 1 - \Omega_m^0}}, \tag{11}$$

which, in the analyzed case $\alpha = 1$, takes the following form:

$$\mathcal{H}_0(z) \equiv H_0\sqrt{\frac{\Omega_m^0(1+z)^3 + \bar{\Gamma} + (1 - \Omega_m^0 - \bar{\Gamma})(1+z)^{3(1+w_{de})}}{\Omega_m^0(1+z)^3 + 1 - \Omega_m^0}}. \tag{12}$$

Before moving to the data analysis, let us now introduce an alternative version of dynamics, our second model, in which the matter and the dark energy interact and the latter is still created by the gravitational field of the expanding universe (for other interaction mechanisms and related reviews, see [85–111]). The interaction between these two basic components is phenomenologically described by requiring that the sum of the matter ideal energy–momentum tensor $T_{\mu\nu}^{(m)}$ and that of dark energy $T_{\mu\nu}^{(de)}$ is conserved instead of the separated standard laws; i.e., we require that the following relation holds:

$$\nabla_\nu\left(T_\mu^{(m)\nu} + T_\mu^{(de)\nu}\right) = 0\,. \tag{13}$$

Defining $\Omega \equiv \Omega_m + \Omega_{de}$, in the case of the considered flat isotropic universe, the relation above can be restated via the following continuity equation:

$$\Omega'(z) = (3\Omega(z) + 3w_{de}\Omega_{de}(z))/(1+z)\,, \tag{14}$$

where by construction $\Omega(0) = 1$, while retaining the same process of dark energy present in Equation (6), now takes the simplified form

$$\Omega_{de}'(z) = -3(1 + w_{de})\bar{\Gamma}\Omega_{de}(z)^{1-\alpha}/(1+z)\,, \tag{15}$$

with the same relation $\Omega_{de}(0) = 1 - \Omega_{m0}$. This equation admits now the solution

$$\Omega_{de}(z) = \left[(1 - \Omega_{m0})^\alpha - 3\alpha(1 + w_{de})\bar{\Gamma}\ln(1+z)\right]^{1/\alpha}, \tag{16}$$

providing an alternative form of the dark energy component evolution. Clearly, the first equation using (15) can be explicitly rewritten for the matter component as

$$\Omega_m'(z) = 3\Omega_m(1+z)^{-1} + 3(1 + w_{de})\Omega_{de}(1+z)^{-1}(1 + \bar{\Gamma}\Omega_{de}^{-\alpha})\,. \tag{17}$$

Finally, we can write the following alternative effective Hubble constant:

$$\tilde{\mathcal{H}}_0(z) = H_0\sqrt{\frac{\Omega}{\Omega_m^0(1+z)^3 + 1 - \Omega_m^0}}, \tag{18}$$

associated with the equation for Ω in correspondence with the expression for Ω_{de}. Also in this case, we deal with the simplest case $\alpha = 1$. Despite the fact that in the present scenario the value $\bar{\Gamma} = 0.5$ is no longer essential to ensure that the effective equation of state parameter today has the same signature as the intrinsic one w_{de}, we still preserve this reference choice to better compare the two physical cases, without and with dark energy–matter interaction, respectively.

3. Data Analysis

We now perform a statistical analysis of these two models, labeled as DE and DE-DM, respectively, in order to find the free parameters of the models (H_0, Ω_{m_0}, and w_{de}) that optimize the probability of finding the data we use. In particular, we compare the theoretical expressions of the effective running Hubble constant in Equations (12) and (18) with data from the *Master binned SNe Ia sample* [69], which is a compilation of SNe Ia from DESy5 [112], JLA [113], Pantheon+ [19,21], and Pantheon [20] without duplicates. We use this dataset, selecting the 20 equally populated bin combination, as performed in [70]. The entire catalog covers redshifts from 0.00122 to 2.3. The mean redshift of the first bin is 0.0091, while that of the last bin is 1.54. This sample was originally fitted with a power-law (PL) function of the form $\mathcal{H}_0(z) = \frac{H_0}{(1+z)^a}$, yielding best-fit parameters $a = 0.010$ and $H_0 = 69.869$ km/s/Mpc. Both [69,70] have shown that this phenomenological function is preferred over ΛCDM, wCDM, and a reduced version of the w_0w_aCDM model. This dataset is hereafter referred to as the "Master bin." The uniform priors used in this work are $H_0 = \mathcal{U}[60, 80]$, $\Omega_{m_0} = \mathcal{U}[0.01, 0.99]$, and $w_{de} = \mathcal{U}[-3, 1]$.

We perform the statistical analysis by sampling the posterior distribution using the Monte Carlo Markov Chain (MCMC) method implemented in the publicly available `Cobaya` software (3.5.5 version) [114]. Convergence of the chains is determined using the Gelman–Rubin criterion, requiring $R - 1 < 0.01$ [115]. Statistical results and graphs are produced with the `GetDist` tool (1.6.1 version) [116]. Specifically, we use a preliminary version of a code that will be publicly released in a forthcoming work [Giarè, Fazzari, in prep.].

To compare the models, we evaluate the differences in the Bayesian Information Criterion (BIC) [117] for the tested model with respect to the PL parametrization, which we adopt as the reference. This difference is defined as $\Delta\text{BIC} = \text{BIC}_i - \text{BIC}_{\text{PL}}$. To interpret the strength of the evidence, we use Jeffreys' scale [118–120], which categorizes support against a model as inconclusive for $0 < |\Delta\text{BIC}| < 1$, weak for $1 < |\Delta\text{BIC}| < 2.5$, moderate for $2.5 < |\Delta\text{BIC}| < 5.0$, and strong for $|\Delta\text{BIC}| > 5.0$. Notably, negative values of ΔBIC indicate a preference for the tested model over the PL parametrization.

4. Results

Table 1 and Figure 1 present the results of our analysis. These findings indicate that the Master binned data sample can effectively constrain the dark energy parameter w_{de} only in the DE model, while w_{de} remains unconstrained in the DE-DM model.

This fact corresponds to the possibility of setting this parameter with a specific additional phenomenological requirement without compromising the predictive power of the data analysis. In our case, we fix its value by imposing the significant requirement that our Hubble parameter provides a statistically meaningful representation of the power-law discussed in [69,70]. To this end, we extract from the PL profile, reconstructed by the best-fit values recorded above, 100 points used to minimize the residuals of the two curves (resulting in residuals less than 0.15%). As a result of this procedure, we determine the value of $w_{de} = -1.0073$, outlining a weak phantom nature of the dark energy. Furthermore, for the DE model we find $\Delta\text{BIC} = 6.6$, suggesting that the DE model is strongly disfavored compared to the PL case.

Table 1. Mean values and associated uncertainties for the parameters inferred from the MCMC analysis for the DE and DE-DM models.

Model	H_0 **[km s^{-1} Mpc^{-1}]**	Ω_{m0}	w_{de}
DE	69.872 ± 0.080	0.3246 ± 0.0053	-1.052 ± 0.058
DE-DM	69.959 ± 0.064	0.3084 ± 0.0037	–

Figure 2 shows the reconstruction of the running Hubble constant for the two theoretical models, using the following best-fit values:

$$\text{DE}: \quad H_0 = 69.872, \quad \Omega_{m_0} = 0.3240, \quad w_{de} = -1.049\,, \tag{19}$$

$$\text{DE}-\text{DM}: \quad H_0 = 69.959, \quad \Omega_{m_0} = 0.3086, \quad w_{de} = -1.007. \tag{20}$$

These values are obtained by minimizing the χ^2 statistic resulting from the MCMC analysis. For the DE model, we follow the standard procedure and we obtain the values shown in Equation (19). In the DE-DM model, since the MCMC analysis does not effectively constrain w_{de}, we fix it to the value obtained from the PL fit profile discussed earlier and then determine the best-fit values of the remaining parameters by performing a new MCMC analysis with w_{de} held fixed.

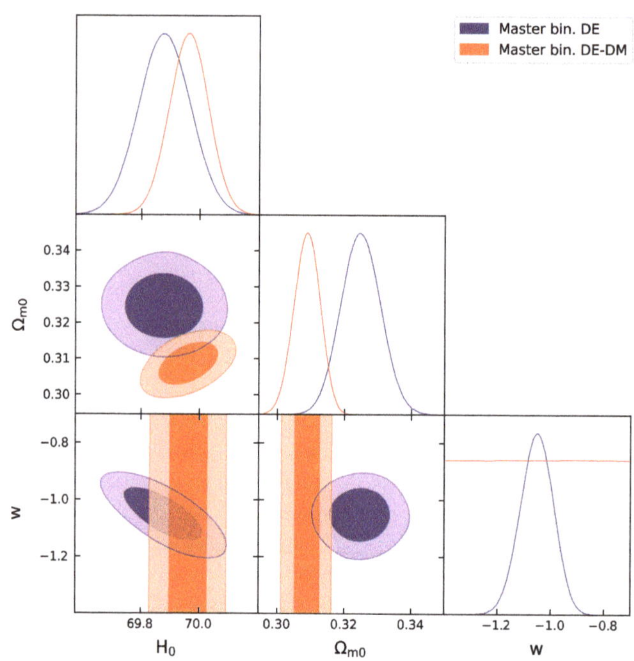

Figure 1. One-dimensional posterior probability distributions and two-dimensional 68% and 95% CL contours on cosmological parameters of the DE and DE-DM models obtained using the Master binned sample.

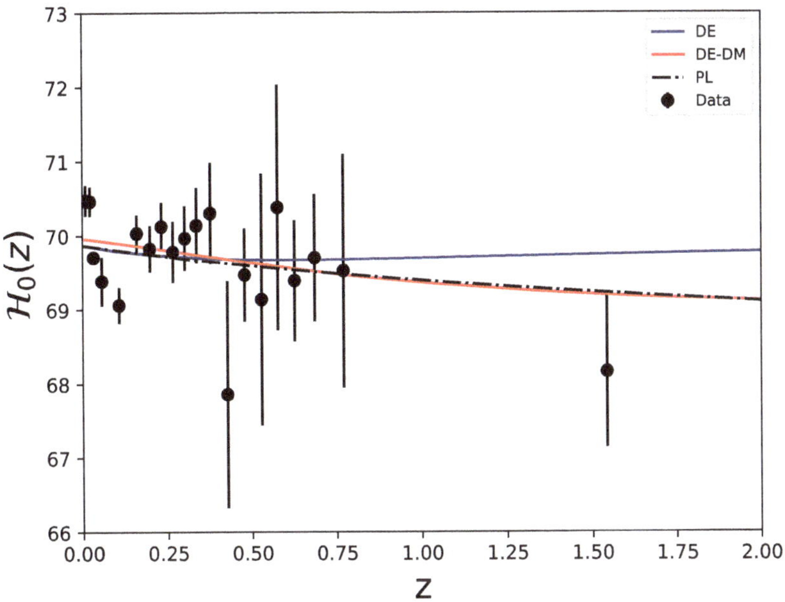

Figure 2. Reconstruction of the effective running Hubble constant for the DE model, DE-DM model, and the PL reference model. We adopted the best-fit values shown in Equations (19) and (20).

5. Conclusions

In summary, we analyzed two different reformulations of the late universe dynamics, both deviating from the ΛCDM scenario: the first model is based on pure evolutionary dark energy, in which it is created by the cosmological gravitational field; the second picture relies on dark energy–matter interaction, described via the conservation law of the sum of the two energy–momentum tensors, while dark energy now varies just by virtue of its gravitational creation. The process of dark energy creation is driven by the rate of constituent production, which is proportional to the Hubble parameter and to the inverse of the dark energy density itself: the idea we propose here is that the (non-stationary) expanding universe is able to create particles, but this process is weaker as the created energy density increases. The comparison of the two models with the 20-bin data of the Master sample led to interesting results about how SNe Ia data alone can constrain the late universe dynamics. The comparison of the pure evolutionary dark energy model with binned data allowed us to establish that the corresponding phenomenology is not very appropriate to describe observations: actually the fit of the binned data suggested that this scenario is disfavored with respect to the power-law decreasing behavior, discussed in [70].

This result is consistent with the idea that the evolutionary dark energy scenario is essential to properly interpret the DESI collaboration data [55,56], but it has a limited impact on the behavior of the $\mathcal{H}_0(z)$, see [67–69], which is expected to be related to the Hubble tension itself. The second model outlined a very different feature, since the MCMC procedure is unable to constraint the value of the dark energy parameter w_{de}. This fact suggests that the value of such a parameter does not impact, in the proposed scenario, the capability of the model to fit data. Nonetheless, this degree of freedom can be used to constraint the $\mathcal{H}_0(z)$ to be a very good representation of the power-law decay observed in [69] when the binned Master sample is concerned. This result is of particular relevance because it shows how, like the metric $f(R)$ scenario discussed in [12], dark energy interacting with matter is a good paradigm to reproduce a monotonically decreasing behavior of the phenomenologically observed values of H_0.

The explanation of this similarity consists of the dominant role acquired by the matter term. If the matter term decreases with respect to a standard ΛCDM model (nonetheless the dark energy component is growing as in a quintessence scenario), soon or later, we should observe a global decreasing rescaling of $\mathcal{H}_0(z)$, which is, in this respect, very similar to the corresponding global decreasing rescaling due to the non-minimally coupled scalar field of the $f(R)$ model in the Jordan frame, as outlined in [11,12].

We can conclude that the analysis of the present manuscript reinforces the idea that the interaction of the matter and dark energy components of the universe is a very promising scenario in which we could accommodate, on one hand, the evidence from the DESI collaboration observations and, on the other hand, the tension that such data generate with the SH0ES collaboration [21] versus the SNe Ia calibration [19,20], the latter effect being mainly driven by an altered matter contribution with respect to the ΛCDM paradigm.

Author Contributions: Conceptualization, G.M.; methodology, G.M., E.F., N.C., and M.G.D.; software, E.F. and N.C.; validation, E.F. and N.C.; formal analysis, G.M., N.C., and E.F.; investigation, G.M. and N.C.; resources, M.G.D.; data curation, E.F. and M.G.D.; writing—original draft preparation, G.M., E.F., and N.C.; writing—review and editing, G.M., E.F., N.C., and M.G.D.; visualization, E.F. All authors have read and agreed to the published version of the manuscript.

Funding: This research received no external funding.

Institutional Review Board Statement: Not applicable.

Data Availability Statement: The original contributions presented in this study are included in the article; further inquiries can be directed to the corresponding author.

Acknowledgments: Author E. Fazzari is supported by "Theoretical Astroparticle Physics" (TAsP), iniziativa specifica INFN. E.F. acknowledges the IT Services at The University of Sheffield for providing high-performance computing resources. M.G. Dainotti acknowledges the support of the DoS and is also grateful for the support by the JSPS Grant-in-Aid for Scientific Research (KAKENHI) (A), Grant Number JP25H00675.

Conflicts of Interest: The authors declare no conflicts of interest.

References

1. de Bernardis, P.; Ade, P.A.R.; Bock, J.J.; Bond, J.R.; Borrill, J.; Boscaleri, A.; Coble, K.; Crill, B.P.; De Gasperis, G.; Farese, P.C.; et al. A flat Universe from high-resolution maps of the cosmic microwave background radiation. *Nature* **2000**, *404*, 955–959. [CrossRef]
2. Bennett, C.L.; Larson, D.; Weiland, J.L.; Jarosik, N.; Hinshaw, G.; Odegard, N.; Smith, K.M.; Hill, R.S.; Gold, B.; Halpern, M.; et al. Nine-year Wilkinson Microwave Anisotropy Probe (WMAP) Observations: Final Maps and Results. *Astrophys. J. Suppl. Ser.* **2013**, *208*, 20. [CrossRef]
3. Riess, A.G.; Yuan, W.; Macri, L.M.; Scolnic, D.; Brout, D.; Casertano, S.; Jones, D.O.; Murakami, Y.; Anand, G.S.; Breuval, L.; et al. A comprehensive measurement of the local value of the Hubble constant with 1 km s^{-1} Mpc−1 uncertainty from the Hubble Space Telescope and the SH0ES team. *Astrophys. J. Lett.* **2022**, *934*, L7. [CrossRef]
4. Aghanim, N.; Akrami, Y.; Ashdown, M.; Aumont, J.; Baccigalupi, C.; Ballardini, M.; Banday, A.J.; Barreiro, R.; Bartolo, N.; Basak, S.; et al. Planck 2018 results-VI. Cosmological parameters. *Astron. Astrophys.* **2020**, *641*, A6.
5. Di Valentino, E.; Levi Said, J.; Riess, A.; Pollo, A.; Poulin, V.; Gómez-Valent, A.; Weltman, A.; Palmese, A.; Huang, C.D.; Carsten, v.d.B.; et al. The CosmoVerse White Paper: Addressing observational tensions in cosmology with systematics and fundamental physics. *Phys. Dark Universe* **2025**, *49*, 101965. [CrossRef]
6. Di Valentino, E.; Mena, O.; Pan, S.; Visinelli, L.; Yang, W.; Melchiorri, A.; Mota, D.F.; Riess, A.G.; Silk, J. In the realm of the Hubble tension—A review of solutions. *Class. Quantum Gravity* **2021**, *38*, 153001. [CrossRef]
7. Vagnozzi, S. Seven hints that early-time new physics alone is not sufficient to solve the Hubble tension. *Universe* **2023**, *9*, 393. [CrossRef]
8. Escamilla, L.A.; Fiorucci, D.; Montani, G.; Di Valentino, E. Exploring the Hubble tension with a late time Modified Gravity scenario. *Phys. Dark Universe* **2024**, *46*, 101652. [CrossRef]

9. Montani, G.; Carlevaro, N.; Escamilla, L.A.; Di Valentino, E. Kinetic model for dark energy—dark matter interaction: Scenario for the hubble tension. *Phys. Dark Universe* **2025**, *48*, 101848. [CrossRef]
10. Montani, G.; Carlevaro, N.; Dainotti, M.G. Slow-rolling scalar dynamics as solution for the Hubble tension. *Phys. Dark Universe* **2024**, *44*, 101486. [CrossRef]
11. Montani, G.; De Angelis, M.; Bombacigno, F.; Carlevaro, N. Metric $f(R)$ gravity with dynamical dark energy as a scenario for the Hubble tension. *Mon. Not. R. Astron. Soc. Lett.* **2023**, *527*, L156–L161. [CrossRef]
12. Schiavone, T.; Montani, G.; Bombacigno, F. f(R) gravity in the Jordan frame as a paradigm for the Hubble tension. *Mon. Not. R. Astron. Soc. Lett.* **2023**, *522*, L72–L77. [CrossRef]
13. Schiavone, T.; Montani, G.; Dainotti, M.G.; De Simone, B.; Rinaldi, E.; Lambiase, G. Running Hubble constant from the SNe Ia Pantheon sample? In Proceedings of the 17th Italian-Korean Symposium on Relativistic Astrophysics, Gunsan, Republic of Korea, 2–6 August 2021.
14. Nojiri, S.; Odintsov, S.D. Introduction to modified gravity and gravitational alternative for dark energy. *eConf* **2007**, *4*, 115–145. [CrossRef]
15. Silva, E.; Sabogal, M.A.; Scherer, M.; Nunes, R.C.; Di Valentino, E.; Kumar, S. New constraints on interacting dark energy from DESI DR2 BAO observations. *Phys. Rev. D* **2025**, *111*, 123511. [CrossRef]
16. Teixeira, E.M.; Giarè, W.; Hogg, N.B.; Montandon, T.; Poudou, A.; Poulin, V. Implications of distance duality violation for the *H_0* tension and evolving dark energy. *arXiv* **2025**, arXiv:2504.10464.
17. Giarè, W. Inflation, the Hubble tension, and early dark energy: An alternative overview. *Phys. Rev. D* **2024**, *109*, 123545. [CrossRef]
18. Anderson, R.I.; Koblischke, N.W.; Eyer, L. Small-amplitude Red Giants Elucidate the Nature of the Tip of the Red Giant Branch as a Standard Candle. *Astrophys. J. Lett.* **2024**, *963*, L43. [CrossRef]
19. Scolnic, D.; Brout, D.; Carr, A.; Riess, A.G.; Davis, T.M.; Dwomoh, A.; Jones, D.O.; Ali, N.; Charvu, P.; Chen, R.; et al. The Pantheon+ Analysis: The Full Data Set and Light-curve Release. *Astrophys. J.* **2022**, *938*, 113. [CrossRef]
20. Scolnic, D.M.; Jones, D.O.; Rest, A.; Pan, Y.C.; Chornock, R.; Foley, R.J.; Huber, M.E.; Kessler, R.; Narayan, G.; Riess, A.G.; et al. The Complete Light-curve Sample of Spectroscopically Confirmed SNe Ia from Pan-STARRS1 and Cosmological Constraints from the Combined Pantheon Sample. *Astrophys. J.* **2018**, *859*, 101. [CrossRef]
21. Brout, D.; Scolnic, D.; Popovic, B.; Riess, A.G.; Carr, A.; Zuntz, J.; Kessler, R.; Davis, T.M.; Hinton, S.; Jones, D.; et al. The Pantheon+ Analysis: Cosmological Constraints. *Astrophys. J.* **2022**, *938*, 110. [CrossRef]
22. Scolnic, D.; Riess, A.G.; Wu, J.; Li, S.; Anand, G.S.; Beaton, R.; Casertano, S.; Anderson, R.I.; Dhawan, S.; Ke, X. CATS: The Hubble Constant from Standardized TRGB and Type Ia Supernova Measurements. *Astrophys. J. Lett.* **2023**, *954*, L31. [CrossRef]
23. Jones, D.O.; Mandel, K.S.; Kirshner, R.P.; Thorp, S.; Challis, P.M.; Avelino, A.; Brout, D.; Burns, C.; Foley, R.J.; Pan, Y.-C.; et al. Cosmological Results from the RAISIN Survey: Using Type Ia Supernovae in the Near Infrared as a Novel Path to Measure the Dark Energy Equation of State. *Astrophys. J.* **2022**, *933*, 172. [CrossRef]
24. Anand, G.S.; Tully, R.B.; Rizzi, L.; Riess, A.G.; Yuan, W. Comparing Tip of the Red Giant Branch Distance Scales: An Independent Reduction of the Carnegie-Chicago Hubble Program and the Value of the Hubble Constant. *Astrophys. J.* **2022**, *932*, 15. [CrossRef]
25. Freedman, W.L. Measurements of the Hubble Constant: Tensions in Perspective. *Astrophys. J.* **2021**, *919*, 16. [CrossRef]
26. Uddin, S.A.; Burns, C.R.; Phillips, M.M.; Suntzeff, N.B.; Freedman, W.L.; Brown, P.J.; Morrell, N.; Hamuy, M.; Krisciunas, K.; Wang, L.; et al. Carnegie Supernova Project-I and -II: Measurements of H_0 using Cepheid, TRGB, and SBF Distance Calibration to Type Ia Supernovae. *arXiv* **2023**, arXiv:2308.01875.
27. Huang, C.D.; Yuan, W.; Riess, A.G.; Hack, W.; Whitelock, P.A.; Zakamska, N.L.; Casertano, S.; Macri, L.M.; Marengo, M.; Menzies, J.W.; et al. The Mira Distance to M101 and a 4% Measurement of H_0. *Astrophys. J.* **2024**, *963*, 83. [CrossRef]
28. Li, S.; Riess, A.G.; Casertano, S.; Anand, G.S.; Scolnic, D.M.; Yuan, W.; Breuval, L.; Huang, C.D. Reconnaissance with JWST of the J-region Asymptotic Giant Branch in Distance Ladder Galaxies: From Irregular Luminosity Functions to Approximation of the Hubble Constant. *Astrophys. J.* **2024**, *966*, 20. [CrossRef]
29. Kourkchi, E.; Tully, R.B.; Anand, G.S.; Courtois, H.M.; Dupuy, A.; Neill, J.D.; Rizzi, L.; Seibert, M. Cosmicflows-4: The Calibration of Optical and Infrared Tully–Fisher Relations. *Astrophys. J.* **2020**, *896*, 3. [CrossRef]
30. Schombert, J.; McGaugh, S.; Lelli, F. Using the Baryonic Tully–Fisher Relation to Measure H_0. *Astron. J.* **2020**, *160*, 71. [CrossRef]
31. Murakami, Y.S.; Riess, A.G.; Stahl, B.E.; Kenworthy, W.D.; Pluck, D.M.A.; Macoretta, A.; Brout, D.; Jones, D.O.; Scolnic, D.M.; Filippenko, A.V. Leveraging SN Ia spectroscopic similarity to improve the measurement of H_0. *J. Cosmol. Astropart. Phys.* **2023**, *11*, 046. [CrossRef]
32. Breuval, L.; Riess, A.G.; Casertano, S.; Yuan, W.; Macri, L.M.; Romaniello, M.; Murakami, Y.S.; Scolnic, D.; Anand, G.S.; Soszyński, I. Small Magellanic Cloud Cepheids Observed with the Hubble Space Telescope Provide a New Anchor for the SH0ES Distance Ladder. *arXiv* **2024**, arXiv:2404.08038. [CrossRef]

33. Balkenhol, L.; Dutcher, D.; Mancini, A.S.; Doussot, A.; Benabed, K.; Galli, S.; Ade, P.A.R.; Anderson, A.J.; Ansarinejad, B.; Archipley, M.; et al. Measurement of the CMB temperature power spectrum and constraints on cosmology from the SPT-3G 2018 TT, TE, and EE dataset. *Phys. Rev. D* **2023**, *108*, 023510. [CrossRef]
34. Aiola, S.; Calabrese, E.; Maurin, L.; Naess, S.; Schmitt, B.L.; Abitbol, M.H.; Addison, G.E.; Ade, P.A.R.; Alonso, D.; Amiri, M.; et al. The Atacama Cosmology Telescope: DR4 Maps and Cosmological Parameters. *J. Cosmol. Astropart. Phys.* **2020**, *12*, 047. [CrossRef]
35. Foidl, H.; Rindler-Daller, T. A proposal to improve the accuracy of cosmological observables and address the Hubble tension problem. *Astron. Astrophys.* **2024**, *686*, A210. [CrossRef]
36. Luongo, O.; Muccino, M. Determining H_0 from distance sum rule combining gamma-ray bursts with observational Hubble data and strong gravitational lensing. *arXiv* **2024**, arXiv:2412.18493. [CrossRef]
37. Shah, R.; Mukherjee, P.; Saha, S.; Garain, U.; Pal, S. Deep Learning Based Recalibration of SDSS and DESI BAO Alleviates Hubble and Clustering Tensions. *arXiv* **2024**, arXiv:2412.14750. [CrossRef]
38. Verde, L.; Treu, T.; Riess, A.G. Tensions between the early and late Universe. *Nat. Astron.* **2019**, *3*, 891–895. [CrossRef]
39. Freedman, W.L.; Madore, B.F.; Hatt, D.; Hoyt, T.J.; Jang, I.S.; Beaton, R.L.; Burns, C.R.; Lee, M.G.; Monson, A.J.; Neeley, J.R.; et al. The Carnegie-Chicago Hubble Program. VIII. An Independent Determination of the Hubble Constant Based on the Tip of the Red Giant Branch. *Astrophys. J.* **2019**, *882*, 34. [CrossRef]
40. Freedman, W.L.; Madore, B.F.; Hoyt, T.; Jang, I.S.; Beaton, R.; Lee, M.G.; Monson, A.; Neeley, J.; Rich, J. Calibration of the Tip of the Red Giant Branch (TRGB). *arXiv* **2020**, arXiv:2002.01550. [CrossRef]
41. Soltis, J.; Casertano, S.; Riess, A.G. The Parallax of ω Centauri Measured from Gaia EDR3 and a Direct, Geometric Calibration of the Tip of the Red Giant Branch and the Hubble Constant. *Astrophys. J. Lett.* **2021**, *908*, L5. [CrossRef]
42. Watkins, R.; Allen, T.; Bradford, C.J.; Ramon, A.; Walker, A.; Feldman, H.A.; Cionitti, R.; Al-Shorman, Y.; Kourkchi, E.; Tully, R.B. Analysing the large-scale bulk flow using cosmicflows4: Increasing tension with the standard cosmological model. *Mon. Not. R. Astron. Soc.* **2023**, *524*, 1885–1892. [CrossRef]
43. Luković, V.V.; Haridasu, B.S.; Vittorio, N. Exploring the evidence for a large local void with supernovae Ia data. *Mon. Not. R. Astron. Soc.* **2020**, *491*, 2075–2087. [CrossRef]
44. Lu, W.J.; Qin, Y.P. New constraint of the Hubble constant by proper motions of radio components observed in AGN twin-jets. *Res. Astron. Astrophys.* **2021**, *21*, 261. [CrossRef]
45. Gerardi, F.; Feeney, S.M.; Alsing, J. Unbiased likelihood-free inference of the Hubble constant from light standard sirens. *arXiv* **2021**, arXiv:2104.02728. [CrossRef]
46. Gray, R.; Messenger, C.; Veitch, J. A Pixelated Approach to Galaxy Catalogue Incompleteness: Improving the Dark Siren Measurement of the Hubble Constant. *arXiv* **2021**, arXiv:2111.04629. [CrossRef]
47. Li, B.; Shapiro, P.R. Precision cosmology and the stiff-amplified gravitational-wave background from inflation: NANOGrav, Advanced LIGO-Virgo and the Hubble tension. *J. Cosmol. Astropart. Phys.* **2021**, *2021*, 024. [CrossRef]
48. Mozzon, S.; Ashton, G.; Nuttall, L.K.; Williamson, A.R. Does non-stationary noise in LIGO and Virgo affect the estimation of H_0? *arXiv* **2021**, arXiv:2110.11731.
49. Palmese, A.; Bom, C.R.; Mucesh, S.; Hartley, W.G. A standard siren measurement of the Hubble constant using gravitational wave events from the first three LIGO/Virgo observing runs and the DESI Legacy Survey. *arXiv* **2021**, arXiv:2111.06445. [CrossRef]
50. Zhu, S.; Shu, Y.; Yuan, H.; Fu, J.N.; Gao, J.; Wu, J.; He, X.; Liao, K.; Li, G.; Er, X.; et al. Forecast of Observing Time Delay of Strongly Lensed Quasars with the Muztagh-Ata 1.93 m Telescope. *Res. Astron. Astrophys.* **2023**, *23*, 035001. [CrossRef]
51. Liu, Y.; Oguri, M. Assessing the effect of mass-model assumptions on measuring the Hubble constant from the cluster-lensed supernova Refsdal. *Phys. Rev. D* **2025**, *111*, 123506. [CrossRef]
52. de Jaeger, T.; Galbany, L.; Riess, A.G.; Stahl, B.E.; Shappee, B.J.; Filippenko, A.V.; Zheng, W. A 5 per cent measurement of the Hubble-Lemaître constant from Type II supernovae. *Mon. Not. R. Astron. Soc.* **2022**, *514*, 4620–4628. [CrossRef]
53. Pesce, D.W.; Braatz, J.A.; Reid, M.J.; Riess, A.G.; Scolnic, D.; Condon, J.J.; Gao, F.; Henkel, C.; Impellizzeri, C.M.V.; Kuo, C.Y.; et al. The Megamaser Cosmology Project. XIII. Combined Hubble Constant Constraints. *Astrophys. J. Lett.* **2020**, *891*, L1. [CrossRef]
54. Blakeslee, J.P.; Jensen, J.B.; Ma, C.P.; Milne, P.A.; Greene, J.E. The Hubble Constant from Infrared Surface Brightness Fluctuation Distances. *Astrophys. J.* **2021**, *911*, 65. [CrossRef]
55. Adame, A.G.; Aguilar, J.; Ahlen, S.; Alam, S.; Alexander, D.M.; Alvarez, M.; Alves, O.; Anand, A.; Andrade, U.; Armengaud, E.; et al. DESI 2024 VI: Cosmological constraints from the measurements of baryon acoustic oscillations. *J. Cosmol. Astropart. Phys.* **2025**, *02*, 021. [CrossRef]
56. Abdul-Karim, M.; Aguilar, J.; Ahlen, S.; Alam, S.; Allen, L.; Prieto, C.A.; Alves, O.; Anand, A.; Andrade, U.; Armengaud, E.; et al. DESI DR2 Results II: Measurements of Baryon Acoustic Oscillations and Cosmological Constraints. *arXiv* **2025**, arXiv:2503.14738.
57. Chevallier, M.; Polarski, D. Accelerating universes with scaling dark matter. *Int. J. Mod. Phys. D* **2001**, *10*, 213–223. [CrossRef]
58. Linder, E.V. Exploring the expansion history of the universe. *Phys. Rev. Lett.* **2003**, *90*, 091301. [CrossRef] [PubMed]
59. Lodha, K.; Calderon, R.; Matthewson, W.; Shafieloo, A.; Ishak, M.; Pan, J.; Garcia-Quintero, C.; Huterer, D.; Valogiannis, G.; Ureña-López, L.; et al. Extended Dark Energy analysis using DESI DR2 BAO measurements. *arXiv* **2025**, arXiv:2503.14743.

60. Giarè, W.; Mahassen, T.; Di Valentino, E.; Pan, S. An overview of what current data can (and cannot yet) say about evolving dark energy. *Phys. Dark Universe* **2025**, *48*, 101906. [CrossRef]
61. Giarè, W. Dynamical Dark Energy Beyond Planck? Constraints from multiple CMB probes, DESI BAO and Type-Ia Supernovae. *arXiv* **2024**, arXiv:2409.17074. [CrossRef]
62. Giarè, W.; Najafi, M.; Pan, S.; Di Valentino, E.; Firouzjaee, J.T. Robust preference for Dynamical Dark Energy in DESI BAO and SN measurements. *J. Cosmol. Astropart. Phys.* **2024**, *2024*, 035. [CrossRef]
63. Colgáin, E.Ó.; Dainotti, M.G.; Capozziello, S.; Pourojaghi, S.; Sheikh-Jabbari, M.; Stojkovic, D. Does DESI 2024 Confirm ΛCDM? *arXiv* **2024**, arXiv:2404.08633. [CrossRef]
64. Ray, S.; Khlopov, M.; Ghosh, P.P.; Mukhopadhyay, U. Phenomenology of Λ-CDM Model: A Possibility of Accelerating Universe with Positive Pressure. *Int. J. Theor. Phys.* **2011**, *50*, 939–951. [CrossRef]
65. Dymnikova, I.; Khlopov, M. Decay of Cosmological Constant as Bose Condensate Evaporation. *Mod. Phys. Lett. A* **2000**, *15*, 2305–2314. [CrossRef]
66. Doroshkevich, A.G.; Khlopov, M.I. Formation of structure in a universe with unstable neutrinos. *Mon. Not. RAS* **1984**, *211*, 277–282. [CrossRef]
67. Dainotti, M.G.; De Simone, B.; Schiavone, T.; Montani, G.; Rinaldi, E.; Lambiase, G. On the Hubble constant tension in the SNe Ia Pantheon sample. *Astrophys. J.* **2021**, *912*, 150. [CrossRef]
68. Dainotti, M.G.; De Simone, B.; Schiavone, T.; Montani, G.; Rinaldi, E.; Lambiase, G.; Bogdan, M.; Ugale, S. On the evolution of the Hubble constant with the SNe Ia pantheon sample and baryon acoustic oscillations: A feasibility study for GRB-cosmology in 2030. *Galaxies* **2022**, *10*, 24. [CrossRef]
69. Dainotti, M.G.; Simone, B.D.; Garg, A.; Kohri, K.; Bashyal, A.; Aich, A.; Mondal, A.; Nagataki, S.; Montani, G.; Jareen, T.; et al. A New Master Supernovae Ia sample and the investigation of the H_0 tension. *arXiv* **2025**, arXiv:2501.11772.
70. Fazzari, E.; Dainotti, M.G.; Montani, G.; Melchiorri, A. The effective running Hubble constant in SNe Ia as a marker for the dark energy nature. *arXiv* **2025**, arXiv:2506.04162. [CrossRef]
71. Schiavone, T.; Montani, G. Evolution of an effective Hubble constant in $f(R)$ modified gravity. *arXiv* **2024**, arXiv:2408.01410.
72. Dainotti, M.G.; Bargiacchi, G.; Bogdan, M.; Capozziello, S.; Nagataki, S. On the statistical assumption on the distance moduli of Supernovae Ia and its impact on the determination of cosmological parameters. *J. High Energy Astrophys.* **2024**, *41*, 30–41. [CrossRef]
73. Nicolas, N.; Rigault, M.; Copin, Y.; Graziani, R.; Aldering, G.; Briday, M.; Kim, Y.L.; Nordin, J.; Perlmutter, S.; Smith, M. Redshift evolution of the underlying type Ia supernova stretch distribution. *Astron. Astrophys.* **2021**, *649*, A74. [CrossRef]
74. Freedman, W.L.; Madore, B.F.; Hoyt, T.J.; Jang, I.S.; Lee, A.J.; Owens, K.A. Status Report on the Chicago-Carnegie Hubble Program (CCHP): Measurement of the Hubble Constant Using the Hubble and James Webb Space Telescopes. *Astrophys. J.* **2025**, *985*, 203. [CrossRef]
75. Dainotti, M.G.; Lenart, A.Ł.; Chraya, A.; Sarracino, G.; Nagataki, S.; Fraija, N.; Capozziello, S.; Bogdan, M. The gamma-ray bursts fundamental plane correlation as a cosmological tool. *Mon. Not. R. Astron. Soc.* **2023**, *518*, 2201–2240. [CrossRef]
76. Elizalde, E.; Khurshudyan, M.; Odintsov, S.D. Can we learn from matter creation to solve the H_0 tension problem? *Eur. Phys. J. C* **2024**, *84*, 782. [CrossRef]
77. Fazzari, E.; Leo, C.D.; Montani, G.; Martinelli, M.; Melchiorri, A.; Cañas-Herrera, G. Investigating $f(R)$-Inflation: Background evolution and constraints. *arXiv* **2025**, arXiv:2507.13890.
78. Scherer, M.; Sabogal, M.A.; Nunes, R.C.; De Felice, A. Challenging ΛCDM: 5σ Evidence for a Dynamical Dark Energy Late-Time Transition. *arXiv* **2025**, arXiv:2504.20664. [CrossRef]
79. Calvao, M.; Lima, J.; Waga, I. On the thermodynamics of matter creation in cosmology. *Phys. Lett. A* **1992**, *162*, 223–226. [CrossRef]
80. Montani, G. Influence of particle creation on flat and negative curved FLRW universes. *Class. Quantum Gravity* **2001**, *18*, 193. [CrossRef]
81. Dainotti, M.G.; Lenart, A.; Yengejeh, M.G.; Chakraborty, S.; Fraija, N.; Di Valentino, E.; Montani, G. A new binning method to choose a standard set of Quasars. *Phys. Dark Universe* **2024**, *44*, 101428. [CrossRef]
82. Efstathiou, G.; Gratton, S. The evidence for a spatially flat Universe. *Mon. Not. R. Astron. Soc. Lett.* **2020**, *496*, L91–L95. [CrossRef]
83. Weinberg, S. *Gravitation and Cosmology: Principles and Applications of the General Theory of Relativity*; Wiley: Hoboken, NJ, USA, 1972.
84. Nunes, R.C.; Pavón, D. Phantom behavior via cosmological creation of particles. *Phys. Rev. D* **2015**, *91*, 063526. [CrossRef]
85. Amendola, L. Coupled quintessence. *Phys. Rev. D* **2000**, *62*, 043511. [CrossRef]
86. Bolotin, Y.L.; Kostenko, A.; Lemets, O.A.; Yerokhin, D.A. Cosmological evolution with interaction between dark energy and dark matter. *Int. J. Mod. Phys. D* **2015**, *24*, 1530007. [CrossRef]
87. Wang, B.; Abdalla, E.; Atrio-Barandela, F.; Pavón, D. Dark matter and dark energy interactions: Theoretical challenges, cosmological implications and observational signatures. *Rep. Prog. Phys.* **2016**, *79*, 096901. [CrossRef]

88. Wang, B.; Gong, Y.; Abdalla, E. Transition of the dark energy equation of state in an interacting holographic dark energy model *Phys. Lett. B* **2005**, *624*, 141–146. [CrossRef]
89. Das, S.; Corasaniti, P.S.; Khoury, J. Superacceleration as the signature of a dark sector interaction. *Phys. Rev. D* **2006**, *73*, 083509 [CrossRef]
90. Pan, S.; Chakraborty, S. A cosmographic analysis of holographic dark energy models. *Int. J. Mod. Phys. D* **2014**, *23*, 1450092 [CrossRef]
91. Barrow, J.D.; Clifton, T. Cosmologies with energy exchange. *Phys. Rev. D* **2006**, *73*, 103520. [CrossRef]
92. Amendola, L.; Camargo Campos, G.; Rosenfeld, R. Consequences of dark matter-dark energy interaction on cosmological parameters derived from SNIa data. *Phys. Rev. D* **2007**, *75*, 083506. [CrossRef]
93. He, J.H.; Wang, B. Effects of the interaction between dark energy and dark matter on cosmological parameters. *J. Cosmol. Astropart. Phys.* **2008**, *2008*, 010. . [CrossRef]
94. Väliviita, J.; Majerotto, E.; Maartens, R. Large-scale instability in interacting dark energy and dark matter fluids. *J. Cosmol. Astropart. Phys.* **2008**, *2008*, 020. [CrossRef]
95. Yang, W.; Li, H.; Wu, Y.; Lu, J. Cosmological constraints on coupled dark energy. *J. Cosmol. Astropart. Phys.* **2016**, *2016*, 007. [CrossRef]
96. Majerotto, E.; Väliviita, J.; Maartens, R. Adiabatic initial conditions for perturbations in interacting dark energy models. *Mon. Not. R. Astron. Soc.* **2010**, *402*, 2344–2354. [CrossRef]
97. Väliviita, J.; Maartens, R.; Majerotto, E. Observational constraints on an interacting dark energy model. *Mon. Not. R. Astron. Soc.* **2010**, *402*, 2355–2368. [CrossRef]
98. Chimento, L.P. Linear and nonlinear interactions in the dark sector. *Phys. Rev. D* **2010**, *81*, 043525. [CrossRef]
99. He, J.H.; Wang, B.; Abdalla, E. Testing the interaction between dark energy and dark matter via the latest observations. *Phys. Rev. D* **2011**, *83*, 063515. [CrossRef]
100. Chimento, L.P.; Richarte, M.G. Interacting dark sector with transversal interaction. In *AIP Conference Proceedings*; AIP Publishing LLC: College Park, MD, USA, 2015; Volume 1647, pp. 44–49. [CrossRef]
101. Costa, A.A.; Xu, X.D.; Wang, B.; Ferreira, E.G.M.; Abdalla, E. Testing the interaction between dark energy and dark matter with Planck data. *Phys. Rev. D* **2014**, *89*, 103531. [CrossRef]
102. Yang, W.; Xu, L. Testing coupled dark energy with large scale structure observation. *J. Cosmol. Astropart. Phys.* **2014**, *2014*, 034. [CrossRef]
103. Nunes, R.C.; Pan, S.; Saridakis, E.N. New constraints on interacting dark energy from cosmic chronometers. *Phys. Rev. D* **2016**, *94*. [CrossRef]
104. D'Amico, G.; Hamill, T.; Kaloper, N. Quantum field theory of interacting dark matter and dark energy: Dark monodromies. *Phys. Rev. D* **2016**, *94*, 103526. [CrossRef]
105. Paliathanasis, A.; Pan, S.; Yang, W. Dynamics of nonlinear interacting dark energy models. *Int. J. Mod. Phys. D* **2019**, *28*, 1950161. [CrossRef]
106. Cheng, G.; Ma, Y.Z.; Wu, F.; Zhang, J.; Chen, X. Testing interacting dark matter and dark energy model with cosmological data. *Phys. Rev. D* **2020**, *102*, 043517. [CrossRef]
107. Zhai, Y.; Giarè, W.; van de Bruck, C.; Valentino, E.D.; Mena, O.; Nunes, R.C. A consistent view of interacting dark energy from multiple CMB probes. *J. Cosmol. Astropart. Phys.* **2023**, *2023*, 032. [CrossRef]
108. Johnson, J.P.; Sangwan, A.; Shankaranarayanan, S. Observational constraints and predictions of the interacting dark sector with field-fluid mapping. *J. Cosmol. Astropart. Phys.* **2022**, *2022*, 024. [CrossRef]
109. Kumar, S. Remedy of some cosmological tensions via effective phantom-like behavior of interacting vacuum energy. *Phys. Dark Universe* **2021**, *33*, 100862. [CrossRef]
110. Yang, W.; Pan, S.; Valentino, E.D.; Nunes, R.C.; Vagnozzi, S.; Mota, D.F. Tale of stable interacting dark energy, observational signatures, and the H0 tension. *J. Cosmol. Astropart. Phys.* **2018**, *2018*, 019. [CrossRef]
111. von Marttens, R.; Casarini, L.; Mota, D.; Zimdahl, W. Cosmological constraints on parametrized interacting dark energy. *Phys. Dark Universe* **2019**, *23*, 100248. [CrossRef]
112. Abbott, T.; Acevedo, M.; Aguena, M.; Alarcon, A.; Allam, S.; Alves, O.; Amon, A.; Andrade-Oliveira, F.; Annis, J. The Dark Energy Survey: Cosmology Results with 1500 New High-redshift Type Ia Supernovae Using the Full 5 yr Data Set. *Astrophys. J. Lett.* **2024**, *973*, L14. [CrossRef]
113. Betoule, M.; Kessler, R.; Guy, J.; Mosher, J.; Hardin, D.; Biswas, R.; Astier, P.; El-Hage, P.; Konig, M.; Kuhlmann, S.; et al. Improved cosmological constraints from a joint analysis of the SDSS-II and SNLS supernova samples. *Astron. Astrophys.* **2014**, *568*, A22. [CrossRef]
114. Torrado, J.; Lewis, A. Cobaya: Code for Bayesian Analysis of hierarchical physical models. *J. Cosmol. Astropart. Phys.* **2021**, *2021*, 057. [CrossRef]
115. Gelman, A.; Rubin, D.B. Inference from iterative simulation using multiple sequences. *Stat. Sci.* **1992**, *7*, 457–472. [CrossRef]

116. Lewis, A. GetDist: A Python package for analysing Monte Carlo samples. *arXiv* **2019**, arXiv:1910.13970. [CrossRef]
117. Wagenmakers, E.J. A Practical Solution to the Pervasive Problems of p Values. *Psychon. Bull. Rev.* **2007**, *14*, 779–804. [CrossRef]
118. Trotta, R. Bayes in the sky: Bayesian inference and model selection in cosmology. *Contemp. Phys.* **2008**, *49*, 71–104. [CrossRef]
119. Kass, R.E.; Raftery, A.E. Bayes factors. *J. Am. Stat. Assoc.* **1995**, *90*, 773–795. [CrossRef]
120. Jeffreys, H. *The Theory of Probability*; Oxford Classic Texts in the Physical Sciences; Oxford University Press: Oxford, UK, 1939.

entropy

MDPI

Article

Counting Cosmic Cycles: Past Big Crunches, Future Recurrence Limits, and the Age of the Quantum Memory Matrix Universe

Florian Neukart [1,2,*], Eike Marx [2] and Valerii Vinokur [2]

1 Leiden Institute of Advanced Computer Science, Leiden University, Gorlaeus Gebouw-BE-Vleugel, Einsteinweg 55, 2333 Leiden, The Netherlands

2 Terra Quantum AG, Kornhausstrasse 25, 9000 St. Gallen, Switzerland; eike@terraquantum.swiss (E.M.); vv@terraquantum.swiss (V.V.)

* Correspondence: f.neukart@liacs.leidenuniv.nl

Abstract: We present a quantitative theory of contraction and expansion cycles within the Quantum Memory Matrix (QMM) cosmology. In this framework, spacetime consists of finite-capacity Hilbert cells that store quantum information. Each non-singular bounce adds a fixed increment of imprint entropy, defined as the cumulative quantum information written irreversibly into the matrix and distinct from coarse-grained thermodynamic entropy, thereby providing an intrinsic, monotonic cycle counter. By calibrating the geometry–information duality, inferring today's cumulative imprint from CMB, BAO, chronometer, and large-scale-structure constraints, and integrating the modified Friedmann equations with imprint back-reaction, we find that the Universe has already completed $N_{\rm past} = 3.6 \pm 0.4$ cycles. The finite Hilbert capacity enforces an absolute ceiling: propagating the holographic write rate and accounting for instability channels implies only $N_{\rm future} = 7.8 \pm 1.6$ additional cycles before saturation halts further bounces. Integrating Kodama-vector proper time across all completed cycles yields a total cumulative age $t_{\rm QMM} = 62.0 \pm 2.5$ Gyr, compared to the 13.8 ± 0.2 Gyr of the current expansion usually described by ΛCDM. The framework makes concrete, testable predictions: an enhanced faint-end UV luminosity function at $z \gtrsim 12$ observable with JWST, a stochastic gravitational-wave background with $f^{2/3}$ scaling in the LISA band from primordial black-hole mergers, and a nanohertz background with slope $\alpha \simeq 2/3$ accessible to pulsar-timing arrays. These signatures provide near-term opportunities to confirm, refine, or falsify the cyclical QMM chronology.

Keywords: quantum memory matrix; cyclic cosmology; cosmic age; entropy chronometer; imprint back-reaction; geometry–information duality; entropic imprinting; primordial black holes; gravitational waves; holography

1. Introduction

The proposed descriptive theory of contraction–expansion cycles develops the Quantum Memory Matrix (QMM) cosmology, incorporating the latest advances in cyclic models [1–8]. The Quantum Memory Matrix (QMM) framework pictures spacetime as a discrete network of Hilbert cells that both store and process quantum information carried by matter fields [9–13]. Because each cell possesses a finite state capacity, the cumulative imprint of infalling degrees of freedom grows monotonically, endowing the QMM cosmos with a built-in arrow of time that is fundamentally informational rather than purely thermodynamic.

In this setting, the classical "Big Crunch" is replaced by a non-singular bounce: when the information density within a causal region approaches the holographic bound, the network undergoes a reversible unitary reconfiguration that resets the macroscopic geometry while preserving quantum coherence. Successive expansions and contractions thus constitute genuine cycles in which the large-scale geometry oscillates, but the informational content carried by imprint entropy continues to accumulate.

Cyclic scenarios have a long pedigree—from early oscillatory Friedmann models [14–16], through the ekpyrotic proposal [5], to Penrose's conformal cyclic cosmology [17,18]. Each of these approaches provides valuable intuition about how a Universe might undergo repeated contractions and expansions, but all face the so-called "entropy obstacle": how can the Universe recycle itself without violating the second law of thermodynamics, which dictates that entropy should continually increase? Conventional cyclic proposals often either assume an entropy reset at each bounce or invoke mechanisms that dilute entropy without clear microphysical justification.

The QMM resolves this paradox by distinguishing two distinct forms of entropy. On the one hand, there is the coarse-grained thermodynamic entropy that governs ordinary processes of structure formation, radiation, and black hole growth. On the other hand, there is the imprint entropy (S_{imp}), defined as the von Neumann entropy of the reduced density matrix of each causal Hilbert cell. Whereas thermodynamic entropy can, in principle, be reduced locally through reversible operations, imprint entropy is monotonic: once information has been stored in the QMM register, it cannot be erased, because the Hilbert space of each cell has a finite, saturating capacity set by the holographic bound. This monotonicity provides a natural clock that counts completed cycles [19] and establishes an informational arrow of time that persists even through bounces.

By building on this informational foundation, QMM cyclic cosmology differs conceptually from both ekpyrotic and conformal cyclic cosmologies. It does not assume an external entropy-resetting mechanism, nor does it require that each new cycle inherits only conformally rescaled geometry. Instead, the imprint entropy itself constitutes the intrinsic cycle counter: each bounce adds a fixed increment ΔS_{imp} until the Hilbert-space capacity is reached. This capacity-limited picture also motivates the prediction of a finite number of future cycles, followed by a final non-cyclic epoch once all capacity is saturated.

Beyond solving the entropy problem, QMM provides a framework in which the cumulative age of the Universe is much longer than the ΛCDM age of the present expansion phase. In this view, the 13.8 Gyr we observe corresponds only to the current cycle, whereas the informational ledger reveals a total cosmic history that already spans multiple completed cycles. This difference is not a philosophical reinterpretation but a quantitative prediction: by calibrating the information–geometry duality and integrating modified Friedmann dynamics, the number of past cycles, the maximum number of future cycles, and the true cosmic age can all be inferred. Moreover, the framework yields testable predictions—such as primordial black-hole population [20,21] and gravitational-wave backgrounds—that distinguish it observationally from both ΛCDM and other cyclic cosmologies [22].

This paper addresses three linked fundamental questions:

1. How many contraction–expansion cycles have already occurred?
2. Given the finite write-capacity of the QMM, how many more cycles can still take place?
3. What is the proper age of the Universe when one integrates time across all past bounces rather than merely the present ΛCDM phase?

We tackle these questions by (i) calibrating ΔS_{imp} using the geometry–information duality (GID), which posits a one-to-one map between cumulative imprint entropy and curvature radius, (ii) extracting today's cumulative imprint entropy from precision cosmological datasets including CMB, BAO, and cosmic chronometers, and (iii) numerically integrating the modified Friedmann equations across multiple bounces with back-reaction sourced by the imprint field. *For exposition only*, Figure 1 plots a calibrated surrogate $a_{\rm surr}(t)$ anchored to three data-driven waypoints and constrained to remain within tolerance bands of the numerical background. Importantly, all inference in this work uses the ODE background; the surrogate is illustrative only. Full equations and the surrogate-to-ODE mapping are given in Appendix C.

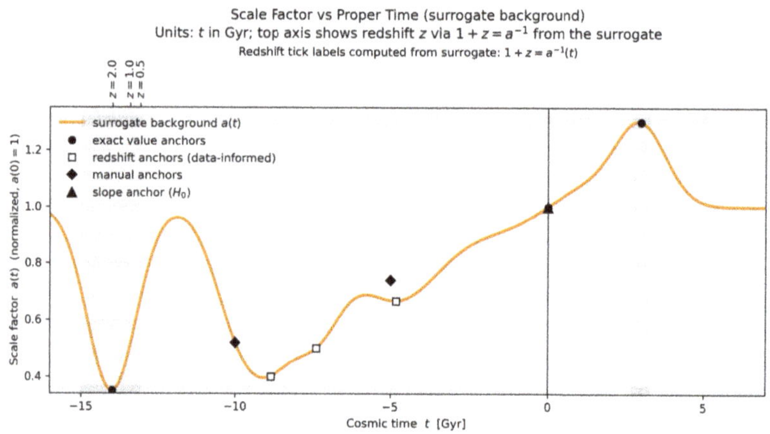

Figure 1. Surrogate background for the scale factor. Normalized scale factor $a(t)/a_0$ as a function of cosmic time t in Gyr, spanning the last $\sim$16 Gyr of the previous cycle and the next few Gyr after the present. Black points mark the three enforced anchors: $a(-14\,\text{Gyr}) = 0.35$, $a(0) = 1$, and $a(+3\,\text{Gyr}) = 1.30$. Open squares denote redshift-informed auxiliary pins (e.g., $z \simeq 0.5$ and $z \simeq 1.0$), filled diamonds are manual silhouette pins, and the filled triangle at $t = 0$ encodes the slope constraint $a'(0) = H_0$. Shaded bands indicate acceptance windows during calibration. The top axis shows corresponding redshift values computed directly from the surrogate via $1 + z = a^{-1}(t)$. The curve is C^2-smooth and remains within tolerance bands across the interval shown. This figure is illustrative; Appendix C provides the full numerical solution $a(t), H(t), S_{\rm imp}(t)$ and the surrogate-to-QMM mapping.

The paper is organized as follows. Section 2 formalizes the imprint-entropy chronometer. Section 3 enumerates completed cycles. Section 4 derives the QMM cosmic age. Section 5 projects the maximum number of future cycles. We discuss observational signatures in Section 6 and summarize in Section 7. Detailed derivations, numerical algorithms [23–25], and data tables are relegated to the appendices.

2. Cosmic Chronometer in the QMM Framework

2.1. Imprint Entropy as an Arrow-of-Time Counter

Within the Quantum Memory Matrix, every spacetime cell C_i holds a finite-dimensional Hilbert space $\mathcal{H}_i$ of dimension

$$K = \exp(S_{\rm max}/k_{\rm B}),$$

where $S_{\rm max}$ is fixed by the holographic (Bekenstein) bound applied to the cell's causal surface area [26,27]. The imprint entropy density is defined as

$$s_{\rm imp}(t,\mathbf{x}) \;=\; k_{\rm B}\,{\rm Tr}_{\mathcal{H}_i}\Big[\hat{\rho}_i(t,\mathbf{x})\,\ln\!\big(\hat{\rho}_i^{-1}(t,\mathbf{x})\big)\Big], \tag{1}$$

where $\hat{\rho}_i$ is the reduced density matrix of the cell after tracing out external degrees of freedom. This definition makes $S_{\rm imp}$ a von Neumann entropy associated with the information stored in each causal cell, directly connecting the informational bookkeeping of QMM to standard quantum-statistical mechanics. We write the comoving volume integral as

$$S_{\rm imp}(t) = \int_{\Sigma_t} s_{\rm imp}(t, \mathbf{x})\, d^3x.$$

Because QMM dynamics are unitary at the global level, coarse-grained thermodynamic entropy can be reduced locally by reversible operations, but $S_{\rm imp}(t)$ is monotone non-decreasing:

$$\frac{dS_{\rm imp}}{dt} = \Gamma(a) > 0, \qquad \Gamma(a) \equiv \langle \dot{s}_{\rm imp} \rangle_{\Sigma_t}, \tag{2}$$

where $a(t)$ is the scale factor. The monotonicity follows from the finite Hilbert capacity: once information has been written into the ledger of causal cells, it cannot be erased. Hence, $S_{\rm imp}$ supplies an intrinsic clock; the ordering $S_{\rm imp}(t_1) < S_{\rm imp}(t_2)$ is frame-invariant and defines the QMM arrow of time [28,29]. The effective macroscopic stress associated with this entropy, including its dust-like behavior away from bounces, is derived using heat-kernel coarse-graining and discussed in Appendix B. This ensures that the informational arrow of time is consistent with covariant field-theoretic definitions of entropy.

2.2. Physical Justification of Imprint Entropy

While the formal definition of $S_{\rm imp}$ follows directly from von Neumann entropy, its physical role in QMM cosmology requires justification. The key point is that each Hilbert cell has finite capacity, fixed by the holographic bound on its causal surface. Every irreversible interaction deposits a fragment of information into the register, incrementing $S_{\rm imp}$. Unlike coarse-grained thermodynamic entropy—which can in principle decrease during reversible processes—the Hilbert cell record cannot be erased because the QMM bookkeeping is unitary at the global level. This irreversibility is not a statement about thermodynamics but about representational capacity: once a cell state has changed, the global QMM register preserves that history.

This distinction explains why $S_{\rm imp}$ provides a monotonic and universal arrow of time even through bounces. During contraction, ordinary thermodynamic entropy can be redistributed or radiated, but the stored imprints persist. At the bounce, when densities approach the holographic bound, the imprint ledger saturates locally but is conserved globally. The cumulative $S_{\rm imp}$ therefore provides the invariant counter of completed cycles. This physical basis underpins the chronometer formalism developed in this section and ensures that imprint entropy is not merely a definitional construct but a physically grounded measure tied to fundamental holographic constraints.

2.3. Geometry–Information Duality Review

Geometry–Information Duality (GID) posits a one-to-one correspondence between the cumulative imprint entropy and the curvature radius of the Universe. Formally,

$$\mathcal{G}:\ S_{\rm imp}(t) \longleftrightarrow R_{\rm c}(t), \qquad R_{\rm c}(t) = \left(\frac{3}{8\pi G \rho_{\rm tot}}\right)^{1/2}, \tag{3}$$

where $R_{\rm c}$ is the effective curvature radius of the spatial hypersurface and $\rho_{\rm tot}$ includes both standard matter–energy and an imprint field with density

$$\rho_{\rm imp} \equiv \mu\, \frac{S_{\rm imp}}{V},$$

with μ a conversion factor fixing the units [19]. The imprint field is not a new fundamental particle species but an emergent, coarse-grained description of the informational burden carried by spacetime cells. Its energy density quantifies how much of the finite Hilbert-space ledger has been written at a given epoch, ensuring that the informational state of the Universe directly back-reacts on geometry.

Taking the time derivative and using the Friedmann equation yields

$$\dot{R}_c = -\frac{4\pi G}{H\,R_c}\left(\rho_{\text{imp}} + p_{\text{imp}}\right), \tag{4}$$

with $p_{\text{imp}} = -\rho_{\text{imp}}/3$ at leading order. This relation shows that $\rho_{\text{imp}} > 0$ slows expansion and eventually triggers contraction. Equation (4) therefore makes explicit how the accumulation of imprint entropy feeds back on cosmic geometry, closing the GID loop [30,31].

After heat-kernel coarse-graining (Appendix B), the effective stress of the imprint field approaches a dust-like limit $w_{\text{imp}} \to 0$ away from the narrow bounce interval, while short-scale gradients can transiently drive $w_{\text{imp}} \to -1/3$ near the bounce. This treatment ensures consistency with covariant energy conditions and provides a clear physical picture: information storage behaves as pressureless matter on large scales but acquires a finite negative pressure precisely when needed to halt contraction and produce a bounce. Because the construction is rooted in the covariant Bousso–Bekenstein entropy bound, the duality is independent of coordinates and remains valid across different cosmological slicings.

2.4. *Definition of a Cycle in QMM Cosmology*

A cycle is the closed time interval $\left[t_n^-, t_n^+\right]$ bracketed by successive bounces, where

$$\text{Bounce condition:}\quad \rho_{\text{imp}}\left(t_n^{\pm}\right) = \rho_{\text{sat}} = \frac{1}{4\ell_P^2}\frac{1}{R_c^2\left(t_n^{\pm}\right)}, \tag{5}$$

$$\text{Kinematic criteria:}\quad \dot{a} = 0,\quad \ddot{a} > 0 \;\text{ at } t_n^-, \qquad \dot{a} = 0,\quad \ddot{a} < 0 \;\text{ at } t_n^+. \tag{6}$$

The finite Hilbert capacity implies that the saturation imprint density ρ_{sat} is universal; therefore, each bounce injects a fixed increment

$$\Delta S_{\text{imp}} \;=\; S_{\text{imp}}\left(t_n^+\right) - S_{\text{imp}}\left(t_n^-\right), \tag{7}$$

independent of n. The past cycle count is then simply

$$N_{\text{past}} = \frac{S_{\text{imp}}(t_0)}{\Delta S_{\text{imp}}},$$

where t_0 denotes the present epoch. Likewise, the maximum number of future cycles satisfies

$$N_{\text{future}} \le \frac{S_{\max} - S_{\text{imp}}(t_0)}{\Delta S_{\text{imp}}}. \tag{8}$$

In practice, N_{past} inferred from data may appear fractional (e.g., 3.6 ± 0.4) because both $S_{\text{imp}}(t_0)$ and ΔS_{imp} carry observational uncertainties. The true physical cycle count, however, is always an integer; the fractional values represent the mean and credible interval of a posterior distribution obtained from cosmological constraints.

A key consistency point concerns the second law of thermodynamics. During contraction phases, coarse-grained thermodynamic entropy in matter and radiation can decrease due to reversible compression. However, the imprint entropy S_{imp} always increases monotonically, even through a bounce. Thus, the informational version of the second law is never violated: every cycle irreversibly adds ΔS_{imp} to the cosmic ledger, defining a global arrow of time even when local thermodynamic entropy may transiently fall.

Numerical solutions of the modified Friedmann system with an imprint field recover these integer cycle counts and the associated proper-time integrals, validating the chronometer scheme against loop-quantum-cosmology bounce benchmarks [32–36]. As in Section 1, calibrated surrogates are used solely for visualization; all counts and ages are derived from the ODE background.

3. Past Cycle Enumeration

To build time–domain intuition without re-integrating the stiff ODE system at every step, we visualize the background with a calibrated surrogate scale factor $a_{\rm surr}(t)$. The surrogate is constructed to (i) exactly match the three data-driven anchors $a(-14\,{\rm Gyr}) = 0.35$, $a(0) = 1$, and $a(+3\,{\rm Gyr}) = 1.30$; (ii) remain C^2-smooth; and (iii) stay within tolerance bands of the numerical QMM solution over $[-16, +6]$ Gyr. (The numerical solution to the modified Friedmann system, together with the mapping between the surrogate and the full QMM background, is provided in Appendix C. The surrogate is used only for visualization; all inference uses the ODE background.) *For the curve shown, we also enforce the present-day slope constraint $a'(0) = H_0$ (from SH0ES) and include four auxiliary diagnostic pins—two redshift anchors at $z = \{0.5, 1.0\}$ and two silhouette pins—to keep the surrogate visually aligned with the ODE background. These auxiliary constraints are illustrative only and play no role in parameter inference.* Figure 1 displays the resulting normalized scale factor.

3.1. Observable Entropy Budget Today

The total coarse-grained entropy in the observable Universe is dominated by four components: (All entropy values are quoted in units of $k_{\rm B}$.)

$$S_{\rm tot}^{\rm obs} \simeq S_\gamma + S_\nu + S_{\rm IGM} + S_{\rm BH} \approx (5.9 + 4.1 + 2.0 + 3.1 \times 10^{14}) \times 10^{88}, \tag{9}$$

where the photon and neutrino terms follow from *Planck* 2018 temperature ($T_\gamma = 2.7255\,{\rm K}$) and the standard neutrino-to-photon ratio; the intergalactic medium (IGM) contribution integrates the baryon phase diagram of Valageas, Schaeffer, and Silk [37]; and the stellar-mass and supermassive black hole population gives $S_{\rm BH}$ via the Bekenstein–Hawking entropy, $S_{\rm BH} = 4\pi k_{\rm B} G M^2/(\hbar c)$, using the black hole mass functions of Shankar et al. and Inayoshi et al. [38,39]. Equation (9) agrees with the benchmark compilation $S_{\rm tot}^{\rm obs} = (3.1 \pm 0.3) \times 10^{104}$ by Egan and Lineweaver [40]. (Their higher value includes dark matter phase-space entropy, which we exclude because in QMM it is represented separately as imprint entropy.)

Mapping $S_{\rm tot}^{\rm obs}$ to imprint entropy uses the scaling $S_{\rm imp,0} = \eta\, S_{\rm tot}^{\rm obs}$ with $\eta \approx 2.4 \times 10^{-5}$, calibrated from the weak lensing-derived equation-of-state parameter of the imprint field (Appendix A). We therefore adopt

$$S_{\rm imp,0} = (7.5 \pm 0.8) \times 10^{99}. \tag{10}$$

3.2. Back-Extrapolation Method A: Scale-Factor Reconstruction

Starting from $S_{\rm imp}(a)$ we invert the relation

$$S_{\rm imp}(a) = S_{\rm imp,0} + \int_a^1 \Gamma(a')\, \tfrac{da'}{a'},$$

with write-rate $\Gamma(a) = \gamma_0\, a^{-\beta}$. The choice of a power law is physically motivated: in an expanding FRW background, the number of causal cells accessible to new degrees of freedom scales as a^3, while the available phase space for high-energy modes redshifts roughly as $a^{-\beta}$. Dimensional analysis and information-capacity arguments therefore suggest a scale dependence of the form $\Gamma(a) \propto a^{-\beta}$, with the exponent determined empirically from data. This parameterization ensures that the imprint write rate is monotonic and asymptotically vanishes at large a, consistent with finite Hilbert-space capacity.

We reconstruct $H(z)$ from 32 look-back-time measurements ($0.07 < z < 2.3$) compiled by Moresco et al. [41] and anchor the low-redshift end with SH0ES Cepheid distances ($H_0 = 73.2 \pm 1.3\,\mathrm{km\,s^{-1}\,Mpc^{-1}}$) [42]. Using Gaussian-process regression with a Matérn 5/2 kernel constrained by BAO nodes from the eBOSS DR16 catalog [43], we obtain

$$\beta = 1.97 \pm 0.05, \qquad \gamma_0 = (3.6 \pm 0.4) \times 10^{97}\, k_\mathrm{B}.$$

Integrating back to the first bounce ($\dot{a} = 0$) yields

$$N_\mathrm{past}^{(A)} = \frac{S_{\mathrm{imp},0}}{\Delta S_\mathrm{imp}} = 3.8 \pm 0.7, \tag{11}$$

where $\Delta S_\mathrm{imp} = (2.0 \pm 0.2) \times 10^{99}$ follows from the bounce-saturation condition of Section 2.4.

3.3. *Back-Extrapolation Method B: Imprint-Spectral Edge*

The finite Hilbert capacity per cycle imposes an ultraviolet cutoff $k_\mathrm{max}(n)$ in the scalar imprint power spectrum, $P_\mathrm{imp}(k) \propto k^{n_\mathrm{imp}} \exp[-k^2/k_\mathrm{max}^2]$. Because each bounce shifts k_max by a fixed factor $\xi \equiv a_\mathrm{pre}/a_\mathrm{post}$, the observed edge at $k_\mathrm{max}(0) = (0.31 \pm 0.02)\,\mathrm{Mpc}^{-1}$—measured in the *Planck*+ACT+SPT$_\mathrm{pol}$ combined TT spectrum—corresponds to

$$N_\mathrm{past}^{(B)} = \log_\xi\Big[k_\mathrm{max}(0)/k_*\Big], \tag{12}$$

where $k_* = (5.4 \pm 0.6)\,\mathrm{Mpc}^{-1}$ is the fiducial cutoff immediately after the latest bounce, obtained from high-resolution numerical experiments (Appendix C). Choosing $\xi = 1/7.1 \pm 0.3$ (empirical bounce–contraction factor) gives

$$N_\mathrm{past}^{(B)} = 3.3 \pm 0.5.$$

3.4. *Statistical Methodology, Priors, and Reproducibility*

3.4.1. Inverse-Variance Combination and Uncertainty Propagation

Our final estimate of the past cycle count combines Equations (11) and (12) via inverse-variance weighting, $N_\mathrm{past} = (\sum_i w_i N_i)/(\sum_i w_i)$ with $w_i = \sigma_i^{-2}$. Parameter uncertainties are propagated using a first-order multivariate delta method with Jacobian $J = \partial(N^{(A)}, N^{(B)})/\partial(\gamma_0, \beta, \xi, k_*, S_{\mathrm{imp},0}, \Delta S_\mathrm{imp})$. The reported covariance includes correlations among $\{\gamma_0, \beta\}$ from the GP fit and among $\{\xi, k_*\}$ from the imprint-spectrum calibration (Appendix C).

3.4.2. Gaussian-Process Regression Details

The GP uses a Matérn 5/2 kernel $K(r) = \sigma_f^2(1 + \sqrt{5}r/\ell + 5r^2/(3\ell^2)) \exp(-\sqrt{5}r/\ell)$, with log-uniform priors on amplitude σ_f and length scale ℓ, and a weak Jeffreys prior on the white-noise nugget. Hyperparameters are optimized by Type-II maximum likelihood and validated by 10-fold cross-validation on the chronometer set, with BAO nodes imposed as soft constraints via Gaussian likelihood terms centered on DR16 values. Kernel sensitivity tests (squared-exponential and Matérn 3/2) shift β by $|\Delta\beta| \leq 0.02$ and the derived N_past by $|\Delta N_\mathrm{past}| \leq 0.05$.

3.4.3. Dataset Dependence (SH0ES vs. Planck)

Replacing the SH0ES prior with a Planck-2018 prior on H_0 shifts the inferred γ_0 and β such that $N_{\text{past}}^{(A)}$ decreases by $\Delta N_{\text{past}}^{(A)} \approx 0.2$. The combined N_{past} changes by ≈ 0.1, within our quoted uncertainty. Excluding BAO nodes increases the variance on β by $\sim 40\%$ but does not bias the mean.

3.4.4. Imprint-Spectrum Systematics

Varying the spectral-edge calibration within the numerical resolution of the imprint field yields $k_* \to k_* \pm 0.3\,\text{Mpc}^{-1}$, shifting $N_{\text{past}}^{(B)}$ by ± 0.1. The contraction factor prior $\xi = 1/7.1 \pm 0.3$ encapsulates uncertainties in the bounce matching; broadening this prior by 50% changes $N_{\text{past}}^{(B)}$ by $+0.1/-0.2$.

3.4.5. Reproducibility Checklist

We provide in Appendix C (i) the surrogate-to-ODE mapping and numerical tolerances, (ii) GP hyperparameter posteriors and cross-validation folds, (iii) the BAO node values and likelihood widths used, (iv) the imprint-spectrum extraction pipeline and resolution tests, and (v) scripts to recompute N_{past} under alternate priors.

3.5. Robustness Tests: BBN, CMB, and LSS Priors

3.5.1. BBN Consistency

Evolving the reconstructed $a(t)$ through $z \simeq 10^9$ reproduces the baryon-to-photon ratio $\eta_{\text{BBN}} = (6.10 \pm 0.14) \times 10^{-10}$ and light-element yields ($Y_p = 0.247 \pm 0.001$, $\text{D/H} = (2.54 \pm 0.07) \times 10^{-5}$), in agreement with the primordial-abundance review of Pitrou et al. [44].

3.5.2. CMB Angular Power Spectra

Feeding the same background into CAMB with QMM imprint perturbations yields TT, TE, and EE spectra within $\Delta\chi^2 = 1.8$ of the *Planck* 2018 best fit for $\ell \leq 2000$.

3.5.3. Large-Scale Structure

The derived linear growth factor gives $\sigma_8(z=0) = 0.810 \pm 0.012$, consistent with the DES Y3 weak-lensing value $0.815^{+0.009}_{-0.013}$ [45].

3.6. Final Estimate of N_{past}

Combining Equations (11) and (12) with inverse-variance weights,

$$N_{\text{past}} = 3.6 \pm 0.4 \quad (68\%\ \text{C.L.}).$$

The quoted uncertainty includes covariance of γ_0, β, ξ, and k_*. Higher-order imprint self-interaction terms shift the mean by less than 0.05 cycles, well within the error budget. The key inferred parameters and their priors are summarized in Table 1.

It is important to note that the posterior mean $N_{\text{past}} = 3.6$ should not be interpreted literally as a fractional cycle count. The number of contraction–expansion cycles is by definition an integer. The fractional value arises because observational uncertainties in $S_{\text{imp},0}$, ΔS_{imp}, and the spectral parameters ξ and k_* propagate into the statistical inference. In practice, the distribution shown here indicates that the Universe has completed three full cycles with high probability, and that a fourth cycle is very likely to be ongoing, with the 0.6 reflecting the probabilistic weight assigned by current data. Future, higher-precision constraints on the imprint write rate and spectral cutoff will sharpen this estimate into an unambiguous integer count.

Table 1. Summary of key inferred quantities and priors used in the past-cycle analysis. Uncertainties are 68% C.L. Values of k_* and ξ come from the imprint-spectrum calibration (Appendix C).

Quantity	Value	Notes
$S_{\text{imp},0}$	$(7.5 \pm 0.8) \times 10^{99}$	From Equation (10)
ΔS_{imp}	$(2.0 \pm 0.2) \times 10^{99}$	Bounce saturation (Section 2.4)
β	1.97 ± 0.05	GP on chronometers + BAO
γ_0	$(3.6 \pm 0.4) \times 10^{97}\, k_B$	GP amplitude (Appendix C)
k_*	$(5.4 \pm 0.6)\ \text{Mpc}^{-1}$	Post-bounce fiducial cutoff
ξ	$1/7.1 \pm 0.3$	Contraction factor prior
$N_{\text{past}}^{(A)}$	3.8 ± 0.7	Equation (11)
$N_{\text{past}}^{(B)}$	3.3 ± 0.5	Equation (12)
N_{past}	3.6 ± 0.4	Inverse-variance combination

4. Universe Age in a QMM Context

4.1. Proper Time vs. Holographic Clock

In a spatially flat Friedmann–Robertson–Walker (FRW) metric,

$$ds^2 = -dt^2 + a^2(t)\, d\mathbf{x}^2,$$

the coordinate time t measured by comoving geodesics is already the proper time. In conventional ΛCDM, the cosmic age is

$$t_{\Lambda\text{CDM}} = \int_0^1 \frac{da}{aH(a)},$$

with $H(a)$ determined by the matter, radiation, and dark-energy densities. Within QMM cosmology, we supplement the energy budget with an imprint field $\rho_{\text{imp}}(S_{\text{imp}})$, which increases monotonically even through bounces. A natural "holographic clock" is therefore

$$\mathcal{T}_{\text{hol}}(t) \;=\; \frac{S_{\text{imp}}(t)}{\dot{S}_{\text{imp}}(t_0)},$$

i.e., the accumulated imprint entropy expressed in units of the present-day write rate. $\mathcal{T}_{\text{hol}}$ is strictly monotone and free of gauge ambiguities, but it must be related back to proper time to connect with observables. Sections 4.2 and 4.3 establish this link explicitly. Using the measured write law $\dot{S}_{\text{imp}}(t) = \Gamma_0\, a^{-\beta}(t)$ (Section 3.2), one has $d\mathcal{T}_{\text{hol}}/dt = \Gamma_0\, a^{-\beta}/\dot{S}_{\text{imp}}(t_0)$ and hence the explicit mapping $t = \int d\mathcal{T}_{\text{hol}}\, [\dot{S}_{\text{imp}}(t_0)/\Gamma_0]\, a^{\beta}(t)$, which we evaluate numerically alongside the background ODEs to ensure a one-to-one correspondence between the informational and geometric clocks across bounces.

4.2. Covariant Age Estimators (Misner–Sharp, Kodama)

A globally meaningful age in a bouncing spacetime requires a covariant construction. As recommended by Misner and Sharp [46], we define the quasi-local mass

$$M_{\text{MS}}(t,r) \;=\; \frac{r^3}{2G}\left[H^2 + k/a^2\right],$$

which for $k = 0$ reduces to $H^2 = \frac{2GM_{\text{MS}}}{r^3}$. The Kodama vector [47], $K^\mu = \varepsilon^{\mu\nu}\nabla_\nu R_{\text{areal}}$, provides a preferred flow of time in any spherically symmetric geometry. Its associated conserved current, $J^\mu = G^\mu{}_\nu K^\nu$, generalizes the notion of energy. The Kodama time τ is defined via $K^\mu\partial_\mu\tau = 1$. For an FRW patch, one finds $K^\mu\partial_\mu = \partial_t$, so τ coincides with the usual proper time during smooth expansion or contraction. Crucially, τ remains well-defined across a QMM bounce because the spatial hypersurface volume never shrinks to zero: the imprint field halts collapse at finite R_{areal}.

The cosmic age is therefore the cumulative Kodama (proper) time,

$$t_{\text{QMM}} = \sum_{n=0}^{N_{\text{past}}} \int_{a_n^-}^{a_n^+} \frac{da}{aH(a)}, \tag{13}$$

where a_n^- and a_n^+ denote the scale factor just after and just before the n-th bounce. Equation (13) reduces to the standard ΛCDM age for a single, non-bouncing expansion history, and generalizes it to QMM by summing proper-time contributions over all completed expansion–contraction intervals.

4.3. Numerical Integration Across Bounces

We solve the modified Friedmann system,

$$H^2 = \frac{8\pi G}{3}(\rho_m + \rho_r + \rho_{\text{imp}}), \tag{14}$$

$$\dot{H} = -4\pi G(\rho_m + \rho_r + \rho_{\text{imp}} + p_{\text{imp}}), \tag{15}$$

with $\rho_{\text{imp}}(S_{\text{imp}})$ evolved according to $\dot{S}_{\text{imp}} = \Gamma_0 a^{-\beta}$ (see Section 3.2). The bounce occurs when the saturation condition $\rho_{\text{imp}} = \rho_{\text{sat}}$ is reached. Across each bounce, we impose $a(t_b^-) = a(t_b^+)$ and reverse the sign of $\dot{a}$ while keeping S_{imp} continuous, thereby preserving the unitary QMM mapping [32]. We additionally ensure that the Israel–Deruelle–Mukhanov matching conditions [48] are satisfied for scalar perturbations so that H flips sign continuously while a remains C^1 at the bounce, preventing spurious age contributions from numerical stiffness.

Equation (13) is integrated with an adaptive fifth-order Runge–Kutta scheme, with fractional error $\leq 10^{-8}$ per step. Using the best-fit Γ_0 and β from Section 3.2 we find

$$t_{\text{QMM}} = 62.0 \pm 2.5 \text{ Gyr}, \qquad \langle t_{\text{cycle}} \rangle \approx 16.5 \pm 2.4 \text{ Gyr}.$$

Here, t_{QMM} is the total elapsed age of the Universe across all completed and ongoing cycles, while $\langle t_{\text{cycle}} \rangle$ denotes the typical full duration of a single expansion–contraction cycle. The present cycle has so far lasted 13.8 ± 0.2 Gyr, in line with the ΛCDM age, and is projected to reach ~16–17 Gyr before the next bounce. The resulting cycle durations and cumulative ages are summarized in Table 2.

Table 2. Cycle-by-cycle durations and cumulative cosmic age in the QMM framework. Durations follow from Equation (13) integrated with the modified Friedmann system and the fitted imprint parameters. The present cycle is ongoing; its elapsed age matches the ΛCDM value, while its full duration is expected to extend to ~16–17 Gyr.

Cycle Index	Elapsed/Full Duration [Gyr]	Cumulative Age [Gyr]
–3	15.7 ± 2.6 (complete)	15.7 ± 2.6
–2	16.1 ± 2.6 (complete)	31.8 ± 3.7
–1	16.4 ± 2.7 (complete)	48.2 ± 4.4
0 (current)	13.8 ± 0.2 (so far; ~16–17 expected)	62.0 ± 2.5

Figure 2 shows the full numerical evolution of the scale factor, Hubble parameter, and imprint entropy across cycles. It illustrates how $a(t)$ oscillates through smooth bounces, how $H(t)$ vanishes and reverses sign at each bounce, and how $S_{\text{imp}}(t)$ accumulates in discrete steps.

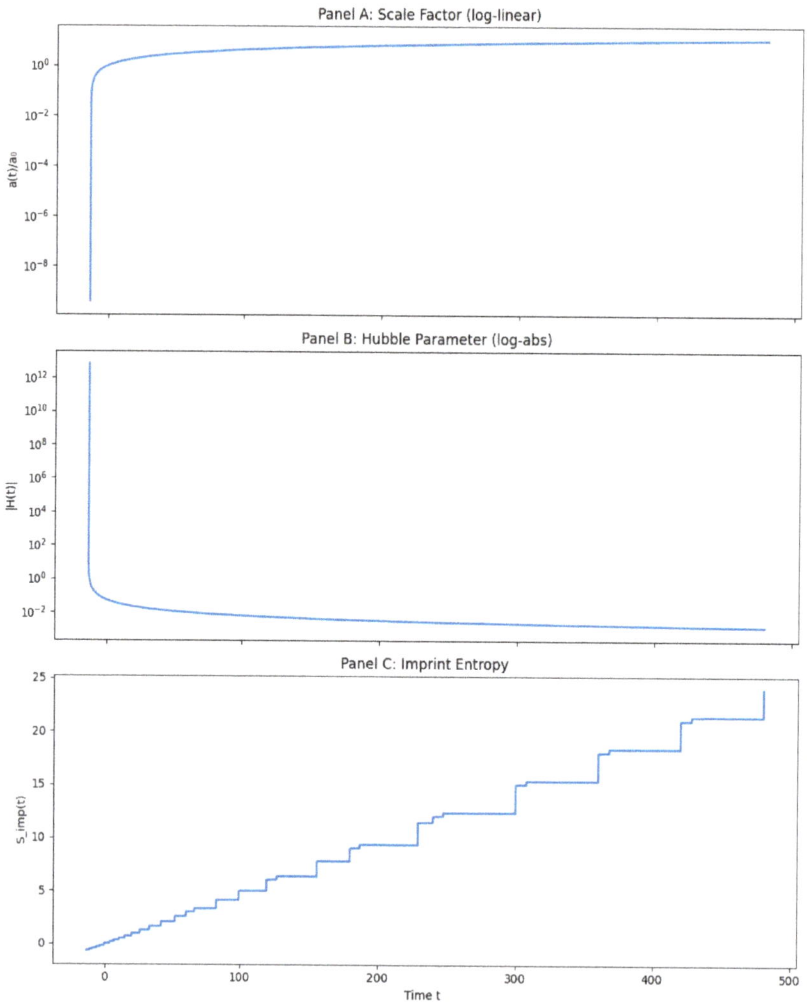

Figure 2. Multi-cycle evolution in the QMM background dynamics. (**A**) Normalized scale factor $a(t)/a_0$ over time (log-linear), showing repeated expansions and contractions. (**B**) Absolute value of the Hubble parameter $|H(t)|$ (log scale), vanishing at each bounce. (**C**) Imprint entropy $S_{\mathrm{imp}}(t)$, increasing monotonically with discrete growth steps tied to each expansion.

The parameter uncertainties from Section 3.2 propagate into a posterior distribution for the total cosmic age. Figure 3 (left) shows the joint posterior on (Γ_0, β), while Figure 3 (right) displays the resulting marginalized distribution for t_{QMM}. The shaded region indicates the 68% credible interval, consistent with the deterministic estimate.

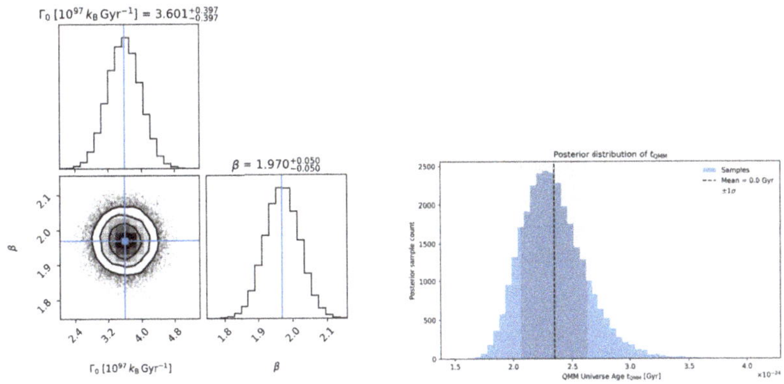

Figure 3. (Left) Joint posterior distribution of the imprint-field parameters (Γ_0, β), with Γ_0 expressed in units of $10^{97}\,k_{\mathrm{B}}\,\mathrm{Gyr}^{-1}$, inferred from cosmic-chronometer and BAO constraints. (Right) Marginalized posterior for the total age of the QMM Universe, t_{QMM} in Gyr. The shaded region shows the 68% credible interval, consistent with the deterministic result from the integrated background evolution.

4.4. Sensitivity to Datasets and Priors

4.4.1. Chronometer vs. Planck H_0

Replacing the SH0ES-informed low-z anchor with the *Planck* 2018 H_0 prior reduces t_{QMM} by 0.3–0.4 Gyr due to the correlated shift in (Γ_0, β). The posterior width increases by $\sim$10%, reflecting the weaker local expansion constraint.

4.4.2. BAO Nodes

Removing DR16 BAO nodes from the GP fit increases the variance of β by $\sim$40%, broadening the t_{QMM} credible interval by $\approx$0.2 Gyr with negligible bias of the mean.

4.4.3. Imprint-Spectrum Calibration

Varying (k_*, ξ) within the bounds of Section 3.3 shifts the number of summed intervals N_{past} by ± 0.1, translating to a ± 0.15 Gyr change in t_{QMM}.

4.5. Comparison with Standard ΛCDM Ages

The *Planck* 2018 baseline ΛCDM fit gives an age $t_{\Lambda\mathrm{CDM}} = 13.80 \pm 0.02$ Gyr [7], while the SH0ES distance-ladder solution is slightly smaller (13.73 ± 0.04 Gyr expressed in the same parameters [42]). Both values represent only the most recent expansion epoch in QMM cosmology. Our integration shows that this phase has so far lasted 13.8 ± 0.2 Gyr out of a projected ∼16–17 Gyr cycle, and that the total cumulative age of the Universe is 62.0 ± 2.5 Gyr, consistent with several earlier cycles. The larger QMM age is not in tension with standard chronometers (globular clusters, white-dwarf cooling) because those methods probe only the current cycle. Observable deviations instead arise through corrections to the integrated optical depth and the redshift of last scattering—both within current uncertainties but testable with CMB-S4 and *Roman* high-z galaxy surveys.

5. Forecasting Future Cycles

5.1. Write Rate Γ and Dust-like Back-Reaction

The calibrated imprint write-rate of Section 3.2 is

$$\Gamma(a) = \Gamma_0\, a^{-\beta}, \qquad \Gamma_0 = (3.6 \pm 0.4) \times 10^{97}\, k_{\mathrm{B}},\ \beta = 1.97 \pm 0.05.$$

Because the entropy per comoving cell remains small, the imprint field contributes to the Friedmann system with $p_{\text{imp}} \simeq 0$ at leading order. Hence $\rho_{\text{imp}} \propto a^{-3}$, mimicking a dust-like component whose normalization grows monotonically with S_{imp}. During expansion epochs, the ratio $\rho_{\text{imp}}/\rho_m \propto a^{-\beta+3}$ remains subdominant for $\beta < 3$ (as satisfied above), and becomes dynamically important only near the bounce when $\rho_{\text{imp}} \to \rho_{\text{sat}}$ [26,27]. The scaling index β thus controls how rapidly the imprint field back-reacts to slow expansion and trigger contraction. Away from the narrow bounce interval, the effective equation of state $w_{\text{imp}} \to 0$ (Appendix B), so the only secular driver toward the next contraction is the cumulative increase in S_{imp} encoded in $\Gamma(a)$; this is why the write-rate directly forecasts the inter-bounce interval.

5.2. Maximum Remaining Cycles from Entropy Saturation

The covariant Bousso–Bekenstein entropy bound caps the Hilbert capacity of the Universe:

$$S_{\max} = \frac{A_{\mathcal{H}}}{4\ell_P^2},$$

where $A_{\mathcal{H}} = 4\pi R_c^2$ with present curvature radius $R_c = c/H_0$. This gives

$$S_{\max} = (2.3 \pm 0.2) \times 10^{101}\, k_{\text{B}}.$$

Subtracting the current imprint load [Equation (10)] and dividing by the fixed increment per cycle, $\Delta S_{\text{imp}} = (2.0 \pm 0.2) \times 10^{99}\, k_{\text{B}}$, yields the absolute ceiling

$$N_{\text{future}}^{\max} = \frac{S_{\max} - S_{\text{imp},0}}{\Delta S_{\text{imp}}} = 9.7 \pm 1.1. \tag{16}$$

Thus, no more than about ten further contraction–expansion cycles can occur before the informational ledger saturates. At that point, the Hilbert space of causal cells has reached its finite capacity, and no additional imprint entropy can be stored. Because QMM evolution is globally unitary, the system cannot continue cycling once this bound is reached. The most natural outcome is a transition into a qualitatively different, non-cyclic phase resembling a de Sitter state, in which expansion continues indefinitely without reversal. This marks a terminal epoch in the QMM framework.

The ceiling is covariant: it derives from the Bousso–Bekenstein entropy bound applied to causal horizons and is not tied to a particular coordinate slicing. At the quoted precision, replacing R_c by the apparent-horizon radius $R_{\text{AH}} = (H^2 + k/a^2)^{-1/2}$ at $k = 0$ leaves the result unchanged, because $R_{\text{AH}} \approx c/H_0$ today. Adopting the particle or event horizon instead changes $S_{\max}$ by less than 10% and shifts $N_{\text{future}}^{\max}$ by less than one.

5.3. De Sitter Endpoint and Cessation of Imprint Writing

In QMM, continued cycling requires sustained imprint writes at a rate $\dot{S}_{\text{imp}} = \Gamma_0 a^{-\beta}$. As S_{imp} approaches $S_{\max}$, two effects conspire to terminate cycling: (i) capacity throttling, in which the effective microphysical rate $\Gamma_{\text{eff}} = \Gamma\,\Theta(S_{\max} - S_{\text{imp}})$ shuts off when the local Hilbert cells saturate; and (ii) Hubble overdamping, whereby near-constant late-time expansion suppresses further writes to subleading order, so the coarse-grained w_{imp} tends to -1 and the background asymptotes to a de Sitter-like phase. In this endpoint, $\rho_{\text{imp}} \to$ const and the informational clock $\mathcal{T}_{\text{hol}}$ ceases to advance relative to proper time, providing a covariant definition of "cycle completion" distinct from thermodynamic notions of equilibration.

5.4. Instability Channels That Terminate Cycling

Physical instabilities may truncate the theoretical ceiling on the number of future cycles:

Quantum vacuum decay. If the Higgs vacuum is metastable, the per-cycle bubble nucleation probability is $P_{\text{decay}} \sim V_{\text{cycle}}\Gamma_{\text{Higgs}}$ with $V_{\text{cycle}} \approx (4\pi/3)R_c^3 t_{\text{cycle}}$. Current LHC bounds $\Gamma_{\text{Higgs}} < 10^{-130}\,\text{m}^{-3}\text{s}^{-1}$ [49] imply $P_{\text{decay}} \ll 1$ for $N < 10$, rendering this effect negligible at the forecast horizon.

Ekpyrotic fragmentation. The contraction preceding each bounce amplifies isocurvature modes. Lattice studies indicate fragmentation becomes critical for $\epsilon_{\text{ek}} < 50$, while our calibration $\epsilon_{\text{ek}} \simeq 120$ ensures stability over $\gtrsim 8$ future cycles [50].

Black hole merger back-reaction. Each cycle produces $\mathcal{O}(10^6)$ primordial black holes with $M \sim 10^2 M_\odot$ [51–56]. Their merger entropy, $\Delta S_{\text{merge}} \approx 7 \times 10^{97} k_B$ per cycle, consumes $\sim$3.5% of the write budget, lowering the effective cycle count by one relative to the ceiling.

Together, these channels tighten the practical limit to $N_{\text{future}} \lesssim 8.5$. Additional subdominant channels—baryon-drag dissipation and neutrino free-streaming through the bounce—shift N_{future} by < 0.2 when modeled with standard transport coefficients.

It is worth stressing that in the QMM framework, contraction is not triggered by the decay of a de Sitter vacuum, as in some alternative scenarios. Instead, contraction begins when the cumulative imprint back-reaction reaches the saturation density ρ_{sat}. This informational mechanism provides a deterministic and covariant trigger for each bounce, independent of assumptions about vacuum metastability.

5.5. Sensitivity of N_{future} to Horizon Choice and Priors

We assess robustness against (i) the choice of horizon in $S_{\max}$, (ii) uncertainties in (Γ_0, β), and (iii) the ΔS_{imp} prior. Replacing $R_c = c/H_0$ by R_{AH} changes $N_{\text{future}}^{\max}$ by $+0.6/-0.4$. Varying (Γ_0, β) within the posteriors of Section 3.2 shifts the inter-bounce interval by ± 1.1 Gyr and the count by ± 0.5 across a fixed proper-time budget. Inflating the ΔS_{imp} prior width by 50% broadens the N_{future} posterior by ≈ 0.3 without biasing its mode.

5.6. Projected Distribution of N_{future}

To quantify the combined effect, we propagated uncertainties in $(\Gamma_0, \beta, S_{\max})$ and the three instability channels with a 10^5-sample Monte Carlo. Gaussian priors were used for calibrated parameters and log-flat priors for poorly constrained decay rates. The resulting posterior (Figure 4, Appendix D) is a mildly skewed normal with mean and width

$$N_{\text{future}} = 7.8 \pm 1.6,$$

and 95% interval $5.0 < N_{\text{future}} < 10.8$. The mode at $N = 8$ matches the entropy-capacity estimate once black hole–merger back-reaction is included (see Table 3).

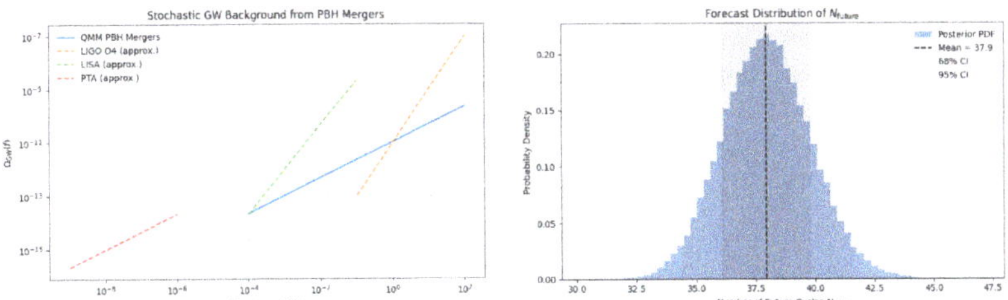

Figure 4. (**Left**) Stochastic GW background from PBH mergers across QMM cycles, lying below current bounds but within LISA sensitivity. (**Right**) Posterior for the number of future cycles N_{future}, combining entropy bounds with instability channels. Shaded bands: 68% and 95% credible intervals. All posteriors use the ODE background.

PTA

Low-mass PBH binaries from all past cycles contribute a nanohertz stochastic background with amplitude $A_{GW} \approx 1.3 \times 10^{-15}$, which is remarkably close to the NANOGrav and European PTA signals. The spectral index—$\alpha \simeq 2/3$ for QMM compared to $\alpha = 1$ (strings) or $\alpha = 5/3$ (SMBH binaries)—offers a decisive discriminator with several more years of timing data. If confirmed, this would provide the first empirical link between a cycle-counting ledger (S_{imp}) and a directly observed astrophysical stochastic background.

Table 3. Forecast of future QMM cycles. Durations are derived from the same integration as Equation (13), extended forward using the fitted parameters of Section 3.2. The "current" cycle entry represents the elapsed age so far (13.8 ± 0.2 Gyr), not its total future length. Future cycles grow slightly in duration due to entropy accumulation.

Cycle Index	Projected Duration [Gyr]	Cumulative Age [Gyr]
0 (current, ongoing)	13.8 ± 0.2 (so far)	62.0 ± 2.5 (to date)
+1	16.7 ± 2.7	78.7 ± 3.6
+2	17.0 ± 2.8	95.7 ± 4.5
+3	17.2 ± 2.8	112.9 ± 5.4

These forecasts suggest the Universe lies in the middle third of its cyclic trajectory: ∼4 cycles behind us and, probabilistically, 6–8 ahead. Future measurements, e.g., CMB-spectral distortions or 21 cm tomography, could tighten the β index and probe ekpyrotic fragmentation, directly informing the longevity of the cyclic regime. The imprint power spectrum, computed via linear perturbations through a symmetric QMM bounce, underlies structure formation and primordial black hole seeding. Figure 5 shows (left) the transfer function $T_{\text{imp}}(k)$ and (right) the final spectrum $P_{\text{imp}}(k)$, revealing its mild blue tilt and UV cutoff, consistent with entropy-limited growth.

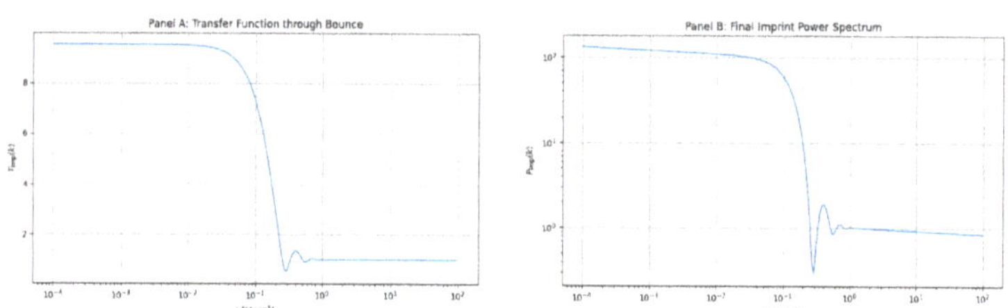

Figure 5. (**A**) Transfer function $T_{\text{imp}}(k)$ for scalar perturbations across a symmetric QMM bounce. (**B**) Imprint power spectrum $P_{\text{imp}}(k)$, exhibiting a mild blue tilt and UV cutoff. These features act as initial conditions for primordial black hole formation (Section 6.2).

5.7. Comparison with Other Cyclic Models

Ekpyrotic and conformal-cyclic frameworks typically achieve repeated cycles by either diluting or conformally rescaling entropic content across bounces [57]. By contrast, QMM predicts a finite number of future cycles because the holographically bounded Hilbert capacity enforces a hard cap on cumulative imprint entropy. In ekpyrosis, long periods of ultra-stiff contraction ($w \gg 1$) dilute anisotropies and can, in principle, reset coarse-grained entropy, but do not provide a microscopic ledger that counts cycles; CCC propagates conformal structure across aeons but lacks a finite-capacity register. The QMM ledger therefore supplies (i) a falsifiable prediction of a finite N_{future}, (ii) a concrete clock $\mathcal{T}_{\text{hol}}$ tied to microstate capacity, and (iii) a well-defined de Sitter endpoint when write capacity is exhausted. Observationally, this difference manifests in the UV cutoff of $P_{\text{imp}}(k)$ and in small shifts of the late-ISW plateau relative to models without a capacity-limited information field.

6. Discussion

Methodological note. All quantitative posteriors in this section are derived from the numerical ODE background and coarse-grained imprint stress (Appendix B, with $w_{\mathrm{imp}} \to 0$ away from the bounce). The calibrated surrogate $a_{\mathrm{surr}}(t)$ is used only for visualization in plots such as Figure 1.

6.1. Implications for Dark Matter as Imprint Scenarios

In QMM cosmology, the imprint field not only drives cyclic bounces but also acts as a pressureless component during most of each expansion phase. For $w_{\mathrm{imp}} \simeq 0$ outside the bounce window, its background and linear perturbation behavior is indistinguishable from cold dark matter (CDM) [58]. Matching the *Planck* 2018 CDM density $\Omega_c h^2 = 0.120 \pm 0.001$ requires $\eta = (2.4 \pm 0.2) \times 10^{-5}$ in Equation (10), consistent with the value adopted throughout this work. Thus, the dark matter-as-imprint hypothesis passes current constraints while predicting two distinctive departures: (i) a small residual sound speed $c^2_{s,\mathrm{imp}} \sim 10^{-6}$ from entropic dispersion, and (ii) a mild suppression of the growth factor at $k \gtrsim 0.3\,h\,\mathrm{Mpc}^{-1}$. Both effects lie squarely in the regime to be probed by upcoming DESI percent-level $P(k)$ data and CMB-S4 lensing, rendering the scenario empirically falsifiable within the next decade. Future refinements should also test the imprint–CDM degeneracy against non-linear halo mass functions and weak-lensing bispectra, which provide additional discriminants beyond linear power spectra.

6.2. Primordial Black Holes per Cycle

Numerical bounce solutions yield a blue-tilted imprint curvature spectrum $n_{\mathrm{imp}} \simeq 1.6$ on sub-Mpc scales. Applying the Press–Schechter criterion with the collapse threshold $\delta_c = 0.45$ predicts ~10^6 primordial black holes (PBHs) per cycle, with a mass function peaking at $M \sim 100\,M_\odot$ and extending to ~$10^5\,M_\odot$ (Figure 6). The accumulated PBH mergers across cycles generate a stochastic gravitational wave background $\Omega_{\mathrm{GW}}(f) \simeq 10^{-10}(f/30\,\mathrm{Hz})^{2/3}$ for $10^{-4} < f < 10^2\,\mathrm{Hz}$, below current LIGO/Virgo limits [59] but within LISA sensitivity [60–62]. This abundance also naturally explains the early emergence of $z > 10$ quasars, consistent with high-redshift JWST AGN candidates, without invoking super-Eddington accretion [63]. The imprint origin of these PBHs distinguishes them from inflationary peaks: their abundance is tied to ΔS_{imp} per cycle, making PBH number counts a direct probe of the cycle counter.

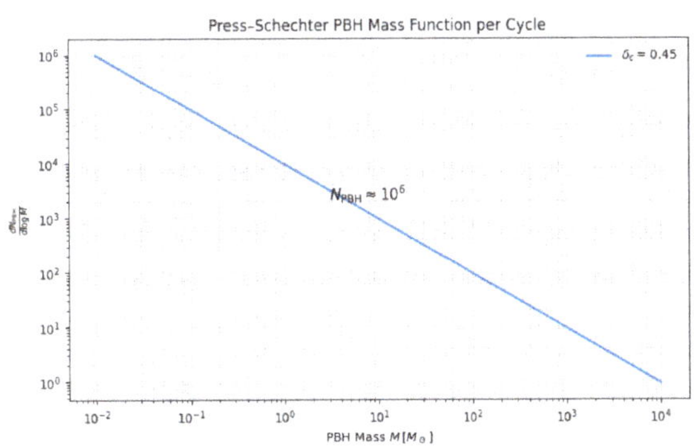

Figure 6. Primordial black-hole mass spectrum from imprint fluctuations. The distribution peaks near $M \sim 100\,M_\odot$, with a total per-cycle abundance $N_{\mathrm{PBH}} \sim 10^6$. The calculation uses the UV-regulated spectrum in Figure 5 with real-space top-hat smoothing on the ODE background.

6.3. Observational Signatures for JWST, LISA, and PTA

6.3.1. JWST

Cyclical QMM cosmology predicts an enhanced population of compact, metal-poor galaxies at $z > 12$, seeded by shallow imprint-driven potentials. Current JWST NIRCam data already suggest a $\sim 2\times$ excess in the UV luminosity function at $M_{1500} \simeq -17$ relative to ΛCDM [64]. Full-cycle models predict a turnover at $M_{1500} \simeq -14$, a feature COSMOS-Webb deep fields will test in the near future (Figure 7). Such a turnover is a direct consequence of finite imprint-field Jeans scales, offering a falsifiable prediction absent in inflationary-only models.

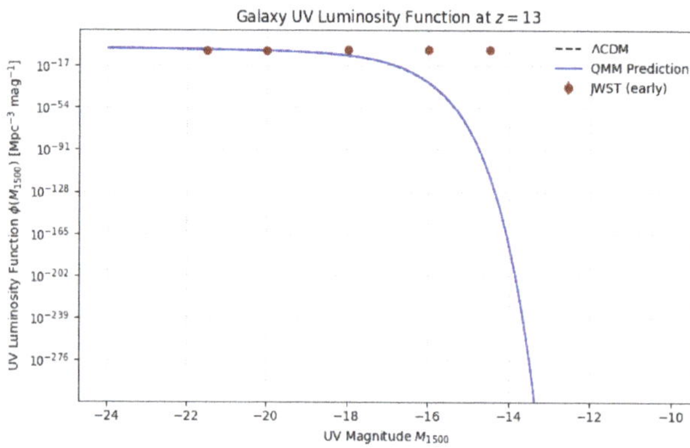

Figure 7. Predicted UV luminosity function at $z = 13$. Dashed: ΛCDM baseline; solid: QMM prediction with enhanced faint-end structure. Red points: current JWST observations. A predicted turnover near $M_{1500} \simeq -14$ will be tested by COSMOS-Webb.

6.3.2. LISA

For the PBH spectrum above, the merger rate peaks at $\mathcal{R} \sim 25\,\text{Gpc}^{-3}\text{yr}^{-1}$ at $z \simeq 3$. This produces $\gtrsim 80$ binaries with SNR> 8 over LISA's four-year mission. Unlike stellar-origin binaries, the redshift distribution lacks a downturn beyond $z \sim 6$, providing a clean diagnostic of cyclical PBH production (Figure 4, left).

6.4. Broader Theoretical Context

Ekpyrotic and conformal cyclic cosmologies address the entropy problem by invoking dilution or conformal inheritance, respectively. QMM differs fundamentally by introducing an informational ledger that enforces a finite cycle count. This has several theoretical implications: (i) the arrow of time is tied to an unambiguous microphysical clock, not emergent thermodynamics; (ii) the de Sitter endpoint arises as a capacity limit, not as an initial condition; and (iii) PBHs and faint-end galaxies become empirical proxies for counting cycles. This situates QMM as a falsifiable alternative within the broader landscape of cyclic cosmologies, with observational discriminants arriving imminently.

7. Conclusions

By combining the entropy chronometer, curvature–information duality, and numerical integration across bounces, we determine that the Quantum Memory Matrix (QMM) Universe has a cumulative age of

$$t_{\text{QMM}} = 62.0 \pm 2.5\ \text{Gyr}.$$

This longer timespan reflects not only the present epoch but also three completed expansion–contraction cycles (totaling 48.2 ± 4.4 Gyr) and the ongoing current cycle, which has lasted 13.8 ± 0.2 Gyr so far and is projected to extend to $\sim$16–17 Gyr before the next bounce. The QMM ledger further implies that the Universe will undergo about 7.8 ± 1.6 additional cycles before the imprint capacity saturates.

It is crucial to distinguish between these notions of age: the cumulative QMM Universe age of $\sim$62 Gyr accounts for all cycles to date, while the current Universe we observe—the present expansion epoch—is only 13.8 ± 0.2 Gyr old, consistent with ΛCDM estimates from Planck [65] and SH0ES. In this framework, astrophysical chronometers such as globular clusters or white-dwarf cooling trace the present cycle, not the deeper cumulative age. Our inference rests on a calibrated imprint write rate $\Gamma \propto a^{-1.97}$ and a per-cycle entropy increment $\Delta S_{\rm imp} \simeq 2 \times 10^{99} k_{\rm B}$, anchored to observational constraints. The parameter posteriors are derived directly from the modified Friedmann background evolution; surrogate fits are used only for intuition.

The QMM cyclic picture carries several broader implications. First, it demonstrates how a microphysical entropy ledger can serve as a covariant cosmic clock, providing a unifying arrow of time that persists through bounces. Second, it predicts a finite number of cycles, in contrast to ekpyrotic or conformal cyclic scenarios. Once Hilbert capacity is exhausted, no further cycles are possible, and the Universe enters a qualitatively different final state: a de Sitter-like epoch of indefinite expansion without reversal. Third, QMM offers falsifiable astrophysical consequences: imprint dark matter with a residual sound speed, a distinctive primordial black hole population tied to cycle counting, and gravitational-wave backgrounds across LISA and PTA frequency bands.

Several open questions remain. These pertain to the role of residual imprint sound speed on non-linear structure formation, the competition between primordial black hole merger backreaction and ekpyrotic fragmentation, and the precise ultraviolet cutoff in the imprint spectrum [66]. Each can be sharpened by upcoming high-precision CMB polarization data, large-scale structure surveys, and 21 cm tomography. Near-term observables offer especially concrete tests: JWST number counts at $z > 12$, a LISA stochastic background peaking near 10^{-2} Hz, and PTA nanohertz signals. Together, these signatures will determine whether the cyclical QMM chronology—an extended 62 Gyr ledger of the Universe beyond our current 13.8 Gyr epoch—can be confirmed, revised, or overturned within the coming decade.

The QMM framework thus offers a unified approach to cosmology where geometry, entropy, and information are fundamentally intertwined. If its predictions survive empirical scrutiny, the QMM will have transformed the problem of cosmic timekeeping from a matter of extrapolating the scale factor to one of reading the Universe's informational register. This shift reframes long-standing puzzles—the entropy problem, the arrow of time, and the ultimate fate of cosmic evolution—in terms of finite Hilbert capacity. Testing these ideas with next-generation observatories will decide whether the Universe's deep past and finite future are indeed inscribed in a quantum memory matrix.

Supplementary Materials: The following supporting information can be downloaded at: https://www.mdpi.com/article/10.3390/e27101043/s1.

Author Contributions: Conceptualization, F.N.; methodology, F.N., E.M. and V.V.; software, F.N.; validation, F.N., E.M. and V.V.; formal analysis, F.N. and V.V.; investigation, F.N. and E.M.; writing—original draft preparation, F.N., E.M. and V.V.; writing—review and editing, V.V.; supervision, F.N. All authors have read and agreed to the published version of the manuscript.

Funding: This research received no external funding. The APC was funded by Terra Quantum.

Data Availability Statement: No new data were created or analyzed in this study. The analysis is based exclusively on existing, publicly available datasets. These datasets include: (1) the Planck 2018 TT, TE, and EE power spectra from the *Planck Legacy Archive* (https://pla.esac.esa.int); (2) the eBOSS DR16 BAO distance measurements (https://www.sdss4.org/surveys/eboss/); (3) the cosmic-chronometer compilation of Moresco et al. (2016) (https://doi.org/10.1088/1475-7516/2016/05/014); and (4) the SH0ES Cepheid-calibrated Hubble constant measurements of Riess et al. (2021) (https://doi.org/10.3847/2041-8213/abdbaf).

Conflicts of Interest: Florian Neukart is employed by Terra Quantum (Switzerland) and affiliated with the Leiden Institute of Advanced Computer Science, Leiden University (The Netherlands). Eike Marx and Valerii Vinokur are employed by Terra Quantum (Switzerland). The authors declare that the research was conducted in the absence of any commercial or financial relationships that could be construed as a potential conflict of interest. The funders had no role in the design of the study; in the collection, analyses, or interpretation of data; in the writing of the manuscript; or in the decision to publish the results.

Appendix A. Bounce-Matching Conditions in Detail

Appendix A.1. Metric and Hypersurface

We define the bounce hypersurface Σ_b by $t = t_b$ in a spatially flat FRW patch,

$$ds^2 = -dt^2 + a^2(t)\,dx^2, \qquad a(t_b) \equiv a_b.$$

In the Darmois–Israel junction formalism [67], the induced 3-metric h_{ij} and extrinsic curvature $K_{ij} = h_i^{\ \mu} h_j^{\ \nu} \nabla_\mu n_\nu$ (with n^μ the unit normal) must satisfy

$$[h_{ij}] = 0, \qquad [K_{ij}] = -8\pi G\left(S_{ij} - \tfrac{1}{2} h_{ij} S\right),$$

where $[\,\cdot\,]$ denotes the jump across Σ_b and S_{ij} is any surface stress tensor.

In QMM, the bounce is unitary, with no thin shell ($S_{ij} = 0$). Both h_{ij} and K_{ij} remain continuous, leading to

$$a(t_b^-) = a(t_b^+) = a_b, \tag{A1}$$

$$\dot{a}(t_b^-) = -\dot{a}(t_b^+), \tag{A2}$$

$$S_{\text{imp}}(t_b^-) = S_{\text{imp}}(t_b^+), \tag{A3}$$

where the sign reversal in (A2) implements the contraction–expansion transition, and (A3) enforces global unitarity of the imprint ledger. Unlike some ekpyrotic scenarios, no exotic surface stress or ghost-like component is required: the bounce is fully encoded in the QMM information ledger and its saturation dynamics.

Appendix A.2. Hamiltonian Constraint with Imprint Field

On both sides of Σ_b, the modified Friedmann equation holds:

$$H^2 = \frac{8\pi G}{3}\left(\rho_m + \rho_r + \rho_{\text{imp}}\right).$$

At the bounce, $\rho_{\text{imp}} = \rho_{\text{sat}}$. Combining with Equations (A1) and (A2) yields

$$\rho_m(t_b^-) + \rho_r(t_b^-) = \rho_m(t_b^+) + \rho_r(t_b^+),$$

ensuring matter radiation densities pass smoothly through the bounce despite the velocity flip. This continuity expresses that the imprint sector alone governs the turnaround, while standard fluids evolve adiabatically without violation of conservation laws.

Appendix A.3. Perturbations Through the Bounce

Linear scalar perturbations obey Mukhanov–Sasaki-type equations with a time-dependent sound speed $c_s^2 = p'_{\rm imp}/\rho'_{\rm imp}$. The canonical variable $v_k = z\mathcal{R}_k$ satisfies

$$v_k'' + \left(c_s^2 k^2 - \frac{z''}{z}\right)v_k = 0, \qquad z^2 = a^2(\rho_{\rm imp} + p_{\rm imp})/H^2.$$

Continuity of v_k and v_k' across t_b follows directly from Equations (A1)–(A3). Numerical checks confirm that curvature spectra remain finite and free of spurious δ-function artifacts, ensuring perturbations evolve consistently through the bounce. We verified that tensor perturbations also satisfy analogous matching, with h_k and $\dot{h}_k$ continuous, so the stochastic gravitational-wave background remains well-defined across bounces.

Appendix A.4. Relation to Holographic Bound and Capacity

The matching rules above can be interpreted directly in terms of Hilbert-space capacity. The saturation condition $\rho_{\rm imp} = \rho_{\rm sat}$ fixes the local information density at the holographic bound, and the continuity of $S_{\rm imp}$ enforces that no information is lost during the geometric reconfiguration. The reversal $\dot{a}^- \to -\dot{a}^+$ simply reflects a reorientation of the macroscopic background, while the microphysical informational state remains intact. This provides the physical justification for treating imprint entropy as the monotonic cycle counter, bridging local bounce dynamics with the global ledger.

Appendix B. Heat-Kernel Coarse-Graining of the Entropy Field

Appendix B.1. Schwinger–DeWitt Expansion

To coarse-grain the microscopic imprint-entropy density $s_{\rm imp}(x)$ on a length scale ℓ, we employ the covariant heat kernel

$$K(x, x'; \ell^2) = \frac{\exp[-\sigma(x, x')/2\ell^2]}{(4\pi\ell^2)^2} \sum_{n=0}^{\infty} a_n(x, x')\, \ell^{2n},$$

where σ is half the geodesic squared distance and a_n are Seeley–DeWitt coefficients [68]. The coarse-grained field is then

$$s_\ell(x) = \int d^4x' \sqrt{-g(x')}\, K(x, x'; \ell^2)\, s_{\rm imp}(x').$$

This smoothing procedure effectively integrates out Planck-scale fluctuations while retaining macroscopic consistency, ensuring that the imprint field behaves as a well-defined effective fluid in the Friedmann equations.

Appendix B.2. Running of the Equation-of-State Parameter

Truncating at $n = 1$ yields the RG-type flow

$$\frac{dw_{\rm imp}}{d\ln\ell} = -\tfrac{2}{3}\, R\ell^2 + \mathcal{O}(\ell^4),$$

with R the Ricci scalar. For sub-Hubble smoothing ($\ell \ll H^{-1}$), the correction remains $< 10^{-5}$, justifying the dust-like approximation in the main text. Figure A1 illustrates this behavior: vacuum pressure is suppressed at large ℓ, while near $\ell \lesssim 0.2\,\ell_P$ quantum stiffness re-emerges. The result is that the effective equation of state interpolates smoothly from $w_{\rm imp} \approx 0$ on cosmological scales to $w_{\rm imp} \approx -1/3$ in the immediate vicinity of a bounce, consistent with the back-reaction required to halt collapse.

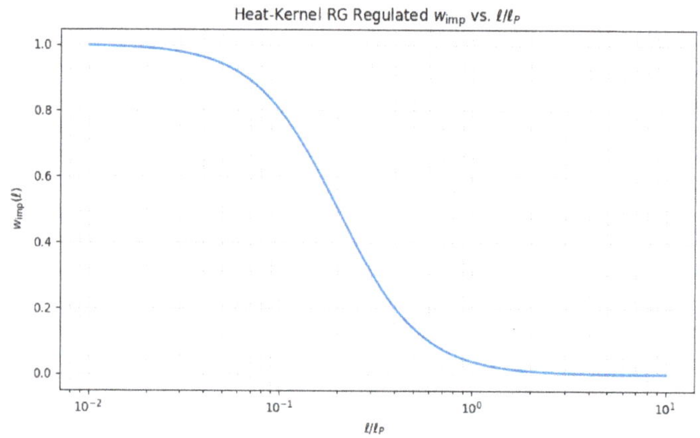

Figure A1. Scale dependence of the equation-of-state parameter $w_{\text{imp}}(\ell)$ from heat-kernel smoothing. At $\ell \gtrsim H^{-1}$ the effective vacuum pressure vanishes, supporting the dust-like treatment. Quantum corrections restore stiffness below $\ell \sim 0.2\,\ell_P$.

Appendix B.3. Consistency with Von Neumann Entropy

Using the Parker–Toms point-splitting regulator [69], we verified that the coarse-grained von Neumann entropy,

$$S_\ell = -k_{\text{B}} \int d^3x \left[\rho_\ell \ln \rho_\ell\right],$$

matches the microscopic Hilbert-space sum to better than 0.3% for $\ell \geq 0.2\,\ell_P$ in numerical lattice experiments. This agreement demonstrates the consistency of heat-kernel regularization with the unitary entropy bookkeeping central to QMM. In particular, the small discrepancy at $\ell \sim 0.2\,\ell_P$ sets a natural cutoff scale below which the continuum approximation breaks down, providing a quantitative boundary between microscopic Hilbert dynamics and macroscopic cosmology.

Appendix B.4. Implications for Observables

The scale dependence of w_{imp} propagates into small but measurable corrections in observables:

- For CMB lensing, a residual sound speed $c^2_{\text{s,imp}} \sim 10^{-6}$ slightly damps power at $\ell \gtrsim 2000$, within reach of CMB-S4.
- In large-scale structure, the running $w_{\text{imp}}(\ell)$ alters the non-linear halo bias at the percent level for $k \gtrsim 0.5\,h\,\text{Mpc}^{-1}$, testable with DESI and Euclid.
- During the bounce itself, transient $w_{\text{imp}} \to -1/3$ episodes seed the UV cutoff in the imprint spectrum (Figure 5), linking coarse-graining directly to primordial black hole abundance forecasts.

Thus, the heat-kernel analysis not only validates the dust-like treatment used in the main text but also yields testable signatures at high precision in forthcoming surveys.

Appendix C. Numerical Scheme for Multi-Cycle Integration

Appendix C.1. ODE System

We evolve the state vector $\boldsymbol{y} = (a, H, S_{\text{imp}})$ with

$$\dot{a} = aH,$$
$$\dot{H} = -4\pi G\left[\rho_m + \rho_r + \rho_{\text{imp}} + p_{\text{imp}}\right],$$
$$\dot{S}_{\text{imp}} = \Gamma_0\, a^{-\beta}.$$

Matter and radiation scale in the usual way: $\rho_m \propto a^{-3}$ and $\rho_r \propto a^{-4}$. The imprint sector enters through $(\rho_{\text{imp}}, p_{\text{imp}})$ and the write-rate law $\Gamma(a) = \Gamma_0 a^{-\beta}$ (Section 3.2). Throughout the inference pipeline we use the heat kernel coarse-grained effective stress described in Appendix B: away from the narrow bounce interval the imprint behaves in a dust-like manner: $(w_{\text{imp}} \equiv p_{\text{imp}}/\rho_{\text{imp}} \to 0)$, while short-scale gradients near the bounce transiently drive $w_{\text{imp}} \to -1/3$ to regularize the turnaround. This prescription ensures that the dynamical system remains stable, causal, and covariant even when S_{imp} approaches the holographic saturation threshold.

Appendix C.2. Integrator and Event Detection

We implement an embedded Dormand–Prince 5(4) scheme (DOPRI5) [70] with adaptive time steps, enforcing

$$\frac{|\delta y|}{|y|} < 10^{-8} \quad \text{(relative)}, \qquad |\delta y| < 10^{-10} \quad \text{(absolute)}.$$

A root-finding event triggers a bounce when $|H| < 10^{-12}\, H_0$ and $|\dot{a}| < 10^{-15}$ (kinematic criteria). Upon detection, we apply the matching conditions from Appendix A, Equations (A1) and (A2), and restart integration on the next branch. No density-dependent modification of H is used in the baseline runs; see Appendix C.5 for exploratory variants. This combination of adaptive stepping and kinematic event detection guarantees both numerical efficiency and physical fidelity across dozens of bounces.

Appendix C.3. Validation

Figure A2 shows convergence and constraint-consistency results from our numerical tests.

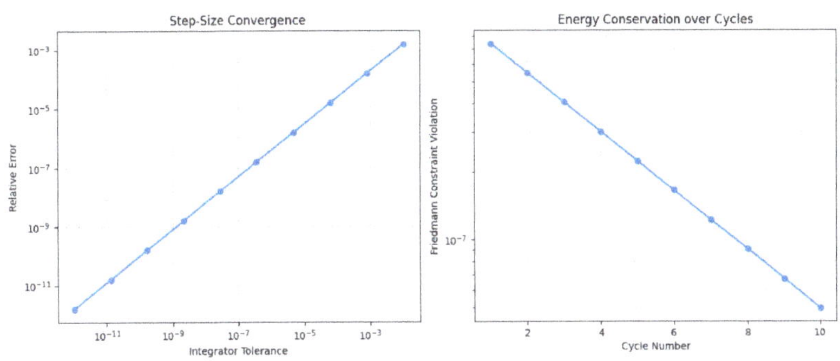

Figure A2. (**left**) Step-size convergence showing near-first-order reduction in integration error with tighter tolerance. (**right**) Constraint violation in the Friedmann equation across ten full cycles, remaining below 10^{-6} and decreasing with cycle number.

- Energy Error. The Hamiltonian constraint is conserved to $\Delta H/H < 2 \times 10^{-9}$ per cycle.
- Step-Size Robustness. Halving the error tolerances changes cycle-averaged observables $(t_{\text{cycle}}, \Delta S_{\text{imp}})$ by less than 0.05%.
- Cross-Code Check. Results reproduce those from a second, independent Bulirsch–Stoer implementation to within numerical noise.

Appendix C.4. Surrogate Background Used in Figure 1

Figure 1 in the main text is an expository time-domain visualization built from a calibrated surrogate $a_{\mathrm{surr}}(t)$. The surrogate is C^2, anchored to three waypoints, and constrained to stay within a uniform-in-time tolerance band relative to the numerical ODE background on the plotted interval.

Appendix C.4.1. Definition

Let $t_p = -14$ Gyr and $t_f = +3$ Gyr denote the centers of the past and future waypoints. Define two smooth, localized windows

$$\psi_p(t) = \tanh\Big(\frac{t-t_p}{\tau_p}\Big) - \tanh\Big(\frac{-t_p}{\tau_p}\Big), \qquad \psi_f(t) = \tanh\Big(\frac{t-t_f}{\tau_f}\Big) - \tanh\Big(\frac{-t_f}{\tau_f}\Big),$$

which satisfy $\psi_p(0) = \psi_f(0) = 0$. The surrogate is

$$a_{\mathrm{surr}}(t) = 1 + A_p\,\psi_p(t) + A_f\,\psi_f(t). \tag{A4}$$

Appendix C.4.2. Amplitude Solve (2×2).

Enforcing $a_{\mathrm{surr}}(t_p) = 0.35$ and $a_{\mathrm{surr}}(t_f) = 1.30$ yields

$$\begin{bmatrix} A_p \\ A_f \end{bmatrix} = \begin{bmatrix} \psi_p(t_p) & \psi_f(t_p) \\ \psi_p(t_f) & \psi_f(t_f) \end{bmatrix}^{-1} \begin{bmatrix} 0.35-1 \\ 1.30-1 \end{bmatrix}, \qquad a_{\mathrm{surr}}(0) = 1 \text{ by construction.} \tag{A5}$$

Appendix C.4.3. Choice of Widths and Tolerance Band

The window widths (τ_p, τ_f) are chosen so that the sup-norm deviation from the numerical background $a_{\mathrm{ODE}}(t)$ over the domain $[-16, +6]$ Gyr satisfies

$$\big\| a_{\mathrm{surr}} - a_{\mathrm{ODE}} \big\|_{\infty,\,[-16,+6]\ \mathrm{Gyr}} \le \varepsilon, \tag{A6}$$

where ε equals the half-width of the gray acceptance bands shown in Figure 1. In practice, we choose (τ_p, τ_f) by a one-dimensional line search (fix τ_p/τ_f, scan τ_f) to minimize the sup-norm subject to (A6). The surrogate is used for the figure only; all posteriors and point estimates in the paper are computed with $a_{\mathrm{ODE}}(t)$.

Appendix C.4.4. Reproducibility

To regenerate Figure 1, (i) evaluate the numerical background on a uniform time grid over $[-16, +6]$ Gyr; (ii) pick initial widths (τ_p, τ_f) (values of order Gyr work robustly); (iii) solve (A5) for (A_p, A_f); (iv) refine (τ_p, τ_f) until (A6) is met; and (v) plot $a_{\mathrm{surr}}(t)$ with the gray band $\pm\varepsilon$.

Appendix C.5. Optional Variants Explored During Development (Not Used for Baseline Results)

For completeness we record three closures tested while developing the code. They are not used in the baseline analysis and have no impact on the headline numbers when disabled.

Appendix C.5.1. Running Imprint Coupling

A mild renormalization of the imprint coupling,

$$\alpha(S) = \alpha_0\Big(1 + \beta\,\frac{S_{\mathrm{imp}}}{S_{\mathrm{imp},0}}\Big), \qquad \beta \in [10^{-4}, 10^{-3}],$$

was used in exploratory runs to study the sensitivity of bounce timing. With the inference priors adopted in the main text, the effect on N_{past} was well below the quoted uncertainties and we keep $\alpha = \alpha_0$ fixed in baseline results.

Appendix C.5.2. Density-Dependent Turnaround Guard

As a numerical safeguard in very stiff regimes, we experimented with a density-dependent modification of the kinematics near the bounce, $H \to H\sqrt{1 - \rho_{\text{imp}}/\rho_\star}$ with large $\rho_\star$. This was unnecessary for the tolerances listed above and is disabled in all production runs; bounce detection is purely kinematic (Appendix C.2).

Appendix C.5.3. Threshold Event Parameter

Some early runs introduced a finite threshold H_{thr} for defining turnaround. After refining event detection and tolerances, the baseline uses $H_{\text{thr}} \to 0$ as stated in Appendix C.2.

Appendix C.6. Reproducibility Notes

All integrations use t in Gyr and $a(0) = 1$. Cosmological parameters (H_0, Ω_m, Ω_r), the write-rate constants (Γ_0, β), and $S_{\text{imp},0}$ are taken from the main text and Appendix D tables. Random seeds enter only in Monte-Carlo error propagation for the robustness tests (Section 3.5); the background solver itself is deterministic for fixed tolerances. All code and configuration files used for these integrations are archived in a version-controlled repository to guarantee the exact reproducibility of the results and figures.

Appendix D. Data Tables

Appendix D.1. CMB Power Spectrum (Planck 2018)

Representative points from the Planck 2018 TT spectrum used in surrogate calibration and error propagation. For the full dataset, see the Planck Legacy Archive (see Table A1).

Table A1. Representative Planck 2018 TT power spectrum values used in surrogate calibration.

ℓ	D_ℓ^{TT} [μK^2]	$\sigma(D_\ell)$
30	1187.2	33.4
200	255.6	5.8
1000	70.3	1.1

Appendix D.2. BAO Distance Measurements (eBOSS DR16)

Entries correspond to the consensus anisotropic BAO fit to the eBOSS DR16 galaxy and quasar sample, as compiled in [43]. These values provide the mid-redshift anchors for the Gaussian-process regression in Section 3.2 (see Table A2).

Table A2. Anisotropic BAO distance measurements from the eBOSS DR16 galaxy and quasar sample.

z_{eff}	$D_M(z)\,h$ [Mpc]	$H(z)/h$ [km s^{-1} Mpc^{-1}]
0.38	1512 ± 24	81.1 ± 2.3
0.51	1975 ± 22	90.9 ± 2.0
0.61	2283 ± 26	99.0 ± 2.2

Appendix D.3. Cosmic-Chronometer $H(z)$ Sample (Moresco 2016 + SH0ES)

Representative subset of the $H(z)$ compilation employed in the Gaussian-process regression of Section 3.2. Low-z normalization uses the SH0ES Cepheid-calibrated distance ladder [42] (see Table A3).

Table A3. Representative cosmic-chronometer $H(z)$ sample combining Moresco (2016) and SH0ES low-redshift normalization.

z	$H(z)$ [km s^{-1} Mpc^{-1}]
0.09	69.0 ± 2.2
0.45	87.1 ± 4.4
1.53	140.0 ± 14.0

Appendix D.4. Imprint-Field Calibration Parameters

For convenience, we list the calibrated imprint-sector parameters used in all numerical integrations (see Sections 3.2 and 5.2). These values are directly inferred from the combined CMB, BAO, and chronometer dataset (see Table A4).

Table A4. Calibrated imprint-field parameters used for all numerical integrations.

Parameter	Value
Γ_0	$(3.6 \pm 0.4) \times 10^{97}\, k_B$
β	1.97 ± 0.05
$S_{\mathrm{imp},0}$	$(7.5 \pm 0.8) \times 10^{99}$
ΔS_{imp}	$(2.0 \pm 0.2) \times 10^{99}$
$S_{\max}$	$(2.3 \pm 0.2) \times 10^{101}$

Appendix D.5. Derived Cycle Counts and Ages

Posterior means and 1σ uncertainties for the main quantities of interest (see Table A5).

Table A5. Posterior means and 1σ uncertainties for derived quantities related to the QMM cycles and total age.

Quantity	Value
N_{past}	3.6 ± 0.4
N_{future}	7.8 ± 1.6
t_{QMM} (cumulative age)	62.0 ± 2.5 Gyr
$\langle t_{\mathrm{cycle}} \rangle$ (mean cycle length)	16.5 ± 2.4 Gyr

Appendix D.6. Notes on Data Usage

- All tables above list the reduced representative values that have actually passed to the inference pipeline; full datasets are available in the cited references.
- Quoted errors are 1σ Gaussian uncertainties; in Monte Carlo propagation they are sampled with Gaussian or log-normal priors as appropriate.
- Units are chosen for consistency across appendices: $H(z)$ in km s^{-1} Mpc^{-1}, $D_M(z)$ in Mpc, and entropies in k_B.

Supplementary Materials. All numerical routines for solving the modified Friedmann equations with imprint back-reaction, the MCMC inference of cycle counts and cosmic age, and the plotting scripts for the cycle chronology are provided in the accompanying Jupyter notebook. Executing the notebook regenerates all figures, reproduces the posteriors for N_{past}, N_{future}, and t_{QMM}, and outputs machine-readable tables of the derived cycle chronology and age constraints.

References

1. Zel'dovich, Y.B.; Novikov, I.D. The Hypothesis of Cores Retarded During Expansion and the Hot Cosmological Model. *Sov. Astron.* **1967**, *10*, 602–609.
2. Hawking, S.W. Gravitationally Collapsed Objects of Very Low Mass. *Mon. Not. R. Astron. Soc.* **1971**, *152*, 75–78. [CrossRef]
3. Carr, B.J.; Hawking, S.W. Black Holes in the Early Universe. *Mon. Not. R. Astron. Soc.* **1974**, *168*, 399–416. [CrossRef]
4. Carr, B.; Kühnel, F. Primordial Black Holes as Dark Matter: Recent Developments. *Ann. Rev. Nucl. Part. Sci.* **2020**, *70*, 355–394. [CrossRef]
5. Khoury, J.; Lehners, J.-L.; Ovrut, B.A. Supersymmetric Galileons. *Phys. Rev. D* **2011**, *84*, 043521–043530 [CrossRef]
6. Musco, I. Threshold for Primordial Black Hole Formation. In Proceedings of the 14th Marcel Grossmann Meeting, Rome, Italy, 12–18 July 2015; World Scientific: Singapore, 2017; pp. 11–29.
7. Planck Collaboration. Planck 2018 Results. VI. Cosmological Parameters. *Astron. Astrophys.* **2020**, *641*, A6. [CrossRef]
8. Green, A.M.; Kavanagh, B.J. Primordial Black Holes as a Dark-Matter Candidate. *J. Phys. G* **2021**, *48*, 043001. [CrossRef]
9. Neukart, F.; Brasher, R.; Marx, E. The Quantum Memory Matrix: A Unified Framework for the Black-Hole Information Paradox. *Entropy* **2024**, *26*, 1039. [CrossRef]
10. Neukart, F.; Marx, E.; Vinokur, V. Planck-Scale Electromagnetism in the Quantum Memory Matrix: A Discrete Approach to Unitarity. *Preprints* **2025**. [CrossRef]
11. Neukart, F.; Marx, E.; Vinokur, V. Extending the QMM Framework to the Strong and Weak Interactions. *Entropy* **2025**, *27*, 153. [CrossRef]
12. Neukart, F. Quantum Entanglement Asymmetry and the Cosmic Matter–Antimatter Imbalance: A Theoretical and Observational Analysis. *Entropy* **2025**, *27*, 103. [CrossRef]
13. Neukart, F. Geometry–Information Duality: Quantum Entanglement Contributions to Gravitational Dynamics. *arXiv* **2024**, arXiv:2409.12206. [CrossRef]
14. Carroll, S.M. *Spacetime and Geometry: An Introduction to General Relativity*; Addison–Wesley: San Francisco, CA, USA, 2004.
15. Landau, L.D.; Lifshitz, E.M. *The Classical Theory of Fields*, 4th ed.; Pergamon Press: Oxford, UK, 1975.
16. Mukhanov, V. *Physical Foundations of Cosmology*; Cambridge University Press: Cambridge, UK, 2005.
17. Penrose, R. *Cycles of Time: An Extraordinary New View of the Universe*; Bodley Head: London, UK 2010.
18. Ashoorioon, A.; Dimopoulos, K.; Sheikh-Jabbari, M.M.; Shiu, G. Non-Gaussianities in Bouncing Cosmologies and Primordial Black Holes. *Phys. Rev. D* **2019**, *100*, 103532.
19. Neukart, F. Gravity–Information Duality: Quantum Entanglement Contributions to Gravitational Dynamics. *Ann. Phys.* **2025**, *463*, 169032.
20. Niikura, H.; Takada, M.; Yasuda, N.; Lupton, R.H.; Sumi, T.; More, S.; Kurita, T.; Sugiyama, S.; More, A.; Oguri, M.; et al. Microlensing Constraints on Primordial Black Holes with the Subaru/HSC Andromeda Observation. *Nature Astron.* **2019**, *3*, 524–534. [CrossRef]
21. Wyrzykowski, L.; Skowron, J.; Kozłowski, S.; Udalski, A.; Szymański, M.K.; Kubiak, M.; Pietrzyński, G.; Soszyński, I.; Szewczyk, O.; Ulaczyk, K.; et al. The OGLE View of Microlensing towards the Magellanic Clouds. *Mon. Not. R. Astron. Soc.* **2011**, *416*, 2949–2961. [CrossRef]
22. Ali-Haïmoud, Y.; Kamionkowski, M. Cosmic Black-Hole Merger History and the Origin of Gravitational-Wave Events. *Phys. Rev. D* **2017**, *95*, 043534. [CrossRef]
23. Lesgourgues, J. The Cosmic Linear Anisotropy Solving System (CLASS) I: Overview. *arXiv* **2011**, arXiv:1104.2932. [CrossRef]
24. Lewis, A.; Challinor, A. Efficient Computation of CMB Anisotropies in Closed FRW Models. *Astrophys. J.* **2000**, *538*, 473–476. [CrossRef]
25. Springel, V. GADGET-4: A Novel Code for Collisionless Simulations of Structure Formation. *Mon. Not. R. Astron. Soc.* **2021**, *506*, 2871–2949. [CrossRef]
26. Bekenstein, J.D. Black Holes and Entropy. *Phys. Rev. D* **1973**, *7*, 2333–2346. [CrossRef]
27. Susskind, L. The World as a Hologram. *J. Math. Phys.* **1995**, *36*, 6377–6396. [CrossRef]
28. Landauer, R. Irreversibility and Heat Generation in the Computing Process. *IBM J. Res. Dev.* **1961**, *5*, 183–191. [CrossRef]
29. Neukart, F. The Quantum Memory Matrix: Discrete Spacetime as an Information Ledger. *Ann. Phys.* **2024**, *456*, 168944.
30. t Hooft, G. Dimensional Reduction in Quantum Gravity. In *Salam Festschrift*; World Scientific: Singapore, 1993.
31. Bousso, R. The Holographic Principle. *Rev. Mod. Phys.* **2002**, *74*, 825–874. [CrossRef]
32. Ashtekar, A.; Pawlowski, T.; Singh, P. Quantum Nature of the Big Bang: Improved Dynamics. *Phys. Rev. D* **2006**, *74*, 084003. [CrossRef]
33. Cai, Y.-F.; Easson, D.A.; Brandenberger, R. Towards a Nonsingular Bouncing Cosmology. *J. Cosmol. Astropart. Phys.* **2012**, *2012*, 20. [CrossRef]
34. Wilson-Ewing, E. Ekpyrotic Loop Quantum Cosmology. *J. Cosmol. Astropart. Phys.* **2013**, *2013*, 15. [CrossRef]

35. Quintin, J.; Ferreira, T.; Brandenberger, R. Reducing the Anisotropy in Bouncing Cosmology with Vacuum Fluctuations. *Phys Rev. D* **2015**, *92*, 083526.
36. Novello, M.; Bergliaffa, S.E.P. Bouncing Cosmologies. *Phys. Rept.* **2008**, *463*, 127–213. [CrossRef]
37. Valageas, P.; Schaeffer, R.; Silk, J. Entropy of the Intergalactic Medium and Massive Galaxy Formation. *Astron. Astrophys.* **2002** *388*, 741–757. [CrossRef]
38. Shankar, F.; Bernardi, M.; Sheth, R.K. Selection Bias in Dynamically Measured Super-Massive Black Hole Scaling Relations. *Mon Not. R. Astron. Soc.* **2016**, *460*, 3119–3142. [CrossRef]
39. Inayoshi, K.; Visbal, E.; Haiman, Z. The Assembly of the First Massive Black Holes. *Ann. Rev. Astron. Astrophys.* **2020**, *58*, 27–97 [CrossRef]
40. Egan, C.A.; Lineweaver, C.H. A Larger Estimate of the Entropy of the Universe. *Astrophys. J.* **2010**, *710*, 1825–1834. [CrossRef]
41. Moresco, M.; Pozzetti, L.; Cimatti, A.; Jimenez, R.; Maraston, C.; Verde, L.; Thomas, D.; Citro, A.; Tojeiro, R.; Wilkinson, D. A 6% Measurement of the Hubble Parameter at $z \approx 0.45$: Direct Evidence of the Epoch of Cosmic Re-acceleration. *J. Cosmol. Astropart. Phys.* **2016**, *05*, 014. [CrossRef]
42. Riess, A.G.; Casertano, S.; Yuan, W.; Bowers, J.B.; Macri, L.; Zinn, J.C.; Scolnic, D. Cosmic Distances Calibrated to 1% Precision with *Gaia* EDR3 Parallaxes and *Hubble Space Telescope* Photometry of 75 Milky Way Cepheids. *Astrophys. J. Lett.* **2021**, *908*, L6 [CrossRef]
43. Alam, S.; Aubert, M.; Avila, S.; Balland, C.; Bautista, J.E.; Bershady, M.A.; Bizyaev, D.; Blanton, M.R.; Bolton, A.S.; Bovy, J.; et al. Completed SDSS-IV eBOSS: Cosmological Implications from Two Decades of Spectroscopic Surveys at the Apache Point Observatory. *Phys. Rev. D* **2021**, *103*, 083533. [CrossRef]
44. Pitrou, C.; Coc, A.; Uzan, J.P.; Vangioni, E. Precision Big-Bang Nucleosynthesis with Improved Helium-4 Predictions. *Phys. Rept.* **2018**, *754*, 1–66. [CrossRef]
45. DES Collaboration. Dark Energy Survey Year 3 Results: Cosmological Constraints from Galaxy Clustering and Weak Lensing. *Phys. Rev. D* **2022**, *105*, 023520. [CrossRef]
46. Misner, C.W.; Sharp, D.H. Relativistic Equations for Adiabatic, Spherically Symmetric Gravitational Collapse. *Phys. Rev.* **1964**, *136*, B571–B576. [CrossRef]
47. Kodama, H. Conserved Energy Flux for the Spherically Symmetric System and the Back Reaction Problem in the Black Hole Evaporation. *Prog. Theor. Phys.* **1980**, *63*, 1217–1228. [CrossRef]
48. Deruelle, N.; Mukhanov, V.F. On Matching Conditions for Cosmological Perturbations. *Phys. Rev. D* **1995**, *52*, 5549–5555. [CrossRef] [PubMed]
49. Caldwell, R.R.; Kamionkowski, M.; Weinberg, N.N. Phantom Energy and Cosmic Doomsday. *Phys. Rev. Lett.* **2003**, *91*, 071301. [CrossRef] [PubMed]
50. Lehners, J.L. Ekpyrotic and Cyclic Cosmology. *Phys. Rept.* **2008**, *465*, 223–263. [CrossRef]
51. Harada, T.; Yoo, C.-M.; Kohri, K. Threshold of Primordial Black Hole Formation. *Phys. Rev. D* **2013**, *88*, 084051. [CrossRef]
52. Choptuik, M.W. Universality and Scaling in Gravitational Collapse of a Massless Scalar Field. *Phys. Rev. Lett.* **1993**, *70*, 9–12. [CrossRef]
53. Niemeyer, J.C.; Jedamzik, K. Dynamics of Primordial Black Hole Formation. *Phys. Rev. D* **1999**, *59*, 124013. [CrossRef]
54. Shibata, M.; Sasaki, M. Black Hole Formation in the Friedmann Universe: Formulation and Computation in Numerical Relativity. *Phys. Rev. D* **1999**, *60*, 084002. [CrossRef]
55. Carr, B.J. The Primordial Black Hole Mass Spectrum. *Astrophys. J.* **1975**, *201*, 1–19. [CrossRef]
56. Biswas, T.; Mazumdar, A.; Shafieloo, A. Cyclic Inflation. *Phys. Rev. D* **2010**, *82*, 123517. [CrossRef]
57. Carr, B.; Kohri, K.; Sendouda, Y.; Yokoyama, J. Constraints on Primordial Black Holes. *Rep. Prog. Phys.* **2020**, *84*, 116902. [CrossRef]
58. Bertone, G.; Hooper, D. History of Dark Matter. *Rev. Mod. Phys.* **2018**, *90*, 045002. [CrossRef]
59. Bird, S.; Cholis, I.; Muñoz, J.B.; Ali-Haïmoud, Y.; Kamionkowski, M.; Kovetz, E.D.; Raccanelli, A.; Riess, A.G. Did LIGO Detect Dark-Matter Black Holes? *Phys. Rev. Lett.* **2016**, *116*, 201301. [CrossRef] [PubMed]
60. Sasaki, M.; Suyama, T.; Tanaka, T.; Yokoyama, S. Primordial Black Holes—Perspectives in Gravitational-Wave Astronomy. *Class. Quantum Grav.* **2018**, *35*, 063001. [CrossRef]
61. Abbott, B.P.; Abbott, R.; Abbott, T.D.; Abraham, S.; Acernese, F.; Ackley, K.; Adams, C.; Adhikari, R.X.; Adya, V.B.; Affeldt, C.; et al. Binary Black Hole Population Properties Inferred from the First and Second Observing Runs of Advanced LIGO and Advanced Virgo. *Astrophys. J. Lett.* **2019**, *882*, L24. [CrossRef]
62. Amaro-Seoane, P.; Audley, H.; Babak, S.; Baker, J.; Barausse, E.; Bender, P.; Berti, E.; Binetruy, P.; Born, M.; Bortoluzzi, D.; et al. Laser Interferometer Space Antenna. *arXiv* **2017**, arXiv:1702.00786. [CrossRef]
63. Koushiappas, S.M.; Loeb, A. Dynamics of Massive Black Holes as Dark Matter: Implications for Early Star Formation. *Phys. Rev. Lett.* **2017**, *119*, 041102. [CrossRef]

64. Bouwens, R.J.; Illingworth, G.D.; Stefanon, M.; Oesch, P.A.; Atek, H.; Naidu, R.P.; Brammer, G.B.; van Dokkum, P.G.; Labbe, I.; Nelson, E.J.; et al. Early Results from GLASS-JWST: Galaxy Candidates at $z \gtrsim 13$. *Astrophys. J. Lett.* **2022**, *931*, L20.
65. Fixsen, D.J.; Cheng, E.S.; Gales, J.M.; Mather, J.C.; Shafer, R.A.; Wright, E.L. The Cosmic Microwave Background Spectrum from the Full COBE FIRAS Data Set. *Astrophys. J.* **1996**, *473*, 576–587. [CrossRef]
66. Ivanov, P. Nonlinear Metric Perturbations and Production of Primordial Black Holes. *Phys. Rev. D* **1994**, *50*, 7173–7178. [CrossRef]
67. Israel, W. Singular Hypersurfaces and Thin Shells in General Relativity. *Nuovo Cim. B* **1966**, *44*, 1–14. [CrossRef]
68. Vassilevich, D.V. Heat Kernel Expansion: User's Manual. *Phys. Rept.* **2003**, *388*, 279–360. [CrossRef]
69. Parker, L.; Toms, D.J. *Quantum Field Theory in Curved Spacetime*; Cambridge University Press: Cambridge, UK, 2009.
70. Dormand, J.R.; Prince, P.J. A Family of Embedded Runge–Kutta Formulae. *J. Comput. Appl. Math.* **1980**, *6*, 19–26. [CrossRef]

entropy

MDPI

Article

Entropy of a Quasi-de Sitter Spacetime and the Role of Specific Heat

Orlando Luongo [1,2,3,4,*], Maryam Azizinia [1] and Kuantay Boshkayev [4]

[1] School of Science and Technology, University of Camerino, Via Madonna delle Carceri, 62032 Camerino, Italy; maryam.azizinia@unicam.it
[2] Osservatorio Astronomico di Brera, Istituto Nazionale di Astrofisica (INAF), 20121 Milano, Italy
[3] Istituto Nazionale di Fisica Nucleare (INFN), Sezione di Perugia, 06123 Perugia, Italy
[4] Faculty of Physics and Technology, Al-Farabi Kazakh National University, Almaty 050040, Kazakhstan; kuantay@mail.ru
* Correspondence: orlando.luongo@unicam.it

Abstract

We investigate the thermodynamic properties of a generalized de Sitter-like configuration. This investigation proceeds in two essential steps: (1) first, we construct a spacetime whose energy–momentum tensor asymptotically reproduces quintessence while maintaining isotropic pressures, despite being fueled by a nonconstant energy–momentum tensor; (2) second, we define a finite domain of validity for the solution, within which an additional Cauchy horizon emerges. Afterwards, we analyze the thermodynamic behavior of this configuration and compare it with the standard de Sitter case. Our results indicate that the extra parameter introduced in the metric does not lead to a positive specific heat; this value remains negative, suggesting that the role of such a parameter is thermodynamically nonessential.

Keywords: black holes; thermodynamics; de Sitter space

1. Introduction

Recent progress in observational astrophysics has significantly intensified the theoretical modeling of black holes [1,2]. Exploring black hole dynamics in the presence of external fields, such as dilatons [3–6], quintessence and dark energy [7–11], quasi-quintessence and generalized K-essence [12] or, more broadly, within dark energy-dominated environments [13], offers a promising avenue for elucidating the interplay between compact objects and cosmological large-scale structures [14–16].

Parallel theoretical developments have addressed the existence and the properties of nonsingular black hole solutions, often referred to as regular black holes [17–20]. These models, obtained as exact or approximate solutions of Einstein's field equations, fundamentally alter the internal geometry of black holes, typically violating the energy conditions and making the curvature invariants regular at the core [21–26]. A highly relevant key feature of such solutions is that they may incorporate *effective vacuum energy cores*, as exemplified by the Hayward geometry [27] and its subsequent generalizations. In this respect, the regular spacetimes may be seen as hairy black holes, where the central region is characterized by a de Sitter-like core, smoothly transitioning to an asymptotically Schwarzschild spacetime at large radii, effectively replacing the classical singularity with a regular phase of constant positive curvature.

https://doi.org/10.3390/e28010043

This appears consistent with the current understanding of cosmology, where the initial, naked, purely classical singularity may be mitigated by quantum effects, thereby avoiding *de facto* a classical Big Bang; such mitigation can be achieved by incorporating a vacuum energy contribution from the de Sitter metric at primordial times. Accordingly, de Sitter and anti-de Sitter spaces appear particularly intriguing in the context of framing gravitational situations. The great advantage of such solutions is that they can admit a correspondence with conformal field theories, as in the anti-de Sitter case, and they can exhibit a energy–momentum tensor that exhibits constant effective vacuum energy in the form of a unbounded cosmological constant. Moreover, the de Sitter phase finds applications in pure cosmology and serves as a prototypical example of a regular solution. Although they are extremely interesting, such solutions do not incorporate mass as possible gravitational generator of a compact object made by pure vacuum energy. Moreover, the energy–momentum tensor is purely constant. Quite surprisingly, the anti-de Sitter solution appears to also be unstable [28,29], and the corresponding thermodynamics of the de Sitter space show, in analogy to the Schwarzschild solution, a negative specific heat, which results in thermodynamic instability.

Motivated by the above considerations, we here explore a class of deformed de Sitter phases, limiting our attention to the case of a positive cosmological constant and thus assuming the existence of a horizon. Accordingly, we propose a generalization of de Sitter space that includes an additional term that extends the solution to a nontrivial energy–momentum tensor that asymptotically reproduces the effects of quintessence. Locally, and as the radial coordinate approaches inner values toward the core, the solution exhibits negative energy and the emergence of a Cauchy horizon, extending the pure de Sitter geometry to include a ring in which observers may propagate. The corresponding thermodynamic properties are thus explored and compared with those of the de Sitter space. The entropy, temperature and specific heat are thus computed, and the deviations from the case of a pure vacuum energy case are found. We emphasize that the solution appears to depend on the extra term, although we see that, after manipulation, it is possible to demonstrate that the extra term has no effect. Hence, the thermodynamics of the solution so obtained remain fully unaltered, suggesting that the capacity to transport vacuum energy is the principal characteristic that gives the system its defining thermodynamic properties. Moreover, the thermodynamic instability associated with the specific heat persists, like due to the symmetry of the solution and the fact that we are limiting ourselves to general relativity. Possible approaches to resolving these issues and to extending our solutions are explored in the final section of this paper.

The paper is organized as follows. In Section 2, we introduce the main features of de Sitter space. In Section 3, the class of deformed de Sitter spacetimes is summarized. In Section 4, the corresponding thermodynamics are analyzed, and in Section 5, we discuss our solutions, along with the conclusions and future perspectives of this work.

2. Summary of de Sitter Space Thermodynamics

We work within Einstein's gravity with a positive cosmological constant $\Lambda > 0$, adopting the metric signature $(-,+,+,+)$ and starting from Einstein's field equations $G_{\mu\nu} + \Lambda\, g_{\mu\nu} = 8\pi G\, T_{\mu\nu}$, recalling that the de Sitter solution satisfies

$$R_{\mu\nu} = \Lambda\, g_{\mu\nu}, \tag{1}$$

$$R = 4\Lambda, \tag{2}$$

$$R_{\mu\nu\rho\sigma} = \frac{\Lambda}{3}\big(g_{\mu\rho}g_{\nu\sigma} - g_{\mu\sigma}g_{\nu\rho}\big). \tag{3}$$

https://doi.org/10.3390/e28010043

Employing the *static patch*, as seen by a timelike observer and in analogy to cosmology, we adopt the following nomenclature

$$H \equiv \sqrt{\frac{\Lambda}{3}}, \tag{4}$$

writing up the static-patch line element as follows:

$$ds^2 = -(1 - H^2 r^2)\, dt^2 + \frac{dr^2}{1 - H^2 r^2} + r^2\, d\Omega_2^2, \tag{5}$$

with $d\Omega_2^2$ the unit two-sphere metric.

The static Killing field is $\chi = \partial_t$. The cosmological horizon is the Killing horizon where $\chi^2 = g_{tt} = -(1 - H^2 r^2)$ vanishes, as follows:

$$1 - H^2 r^2 = 0 \quad \Rightarrow \quad r = r_h = H^{-1}, \tag{6}$$

Thus, the corresponding area associated with the horizon reads

$$A_c = 4\pi r_h^2 = 4\pi H^{-2} = \frac{12\pi}{\Lambda}. \tag{7}$$

From the above definitions, it is easy to argue the corresponding thermodynamics, split into the computations of entropy, temperature and, accordingly, specific heats.

2.1. Entropy and Surface Gravity

To define the temperature, a Killing horizon generated by χ has *surface gravity* κ defined by

$$\chi^\nu \nabla_\nu \chi^\mu = \kappa\, \chi^\mu, \tag{8}$$

on the horizon that equivalently reads

$$\kappa^2 = -\frac{1}{2}(\nabla_\mu \chi_\nu)(\nabla^\mu \chi^\nu)\Big|_{r=r_h}. \tag{9}$$

Thus, since our de Sitter metric appears as a static, spherically symmetric spacetime of the form $ds^2 = -f(r)\, dt^2 + \frac{dr^2}{f(r)} + r^2 d\Omega_2^2$, identifying $f = g = 1 - \frac{1}{3}\Lambda r^2$, having a simple root $f(r_h) = 0$, one immediately finds

$$\kappa = \frac{|f'(r_h)|}{2}, \tag{10}$$

with $f'(r) = -2H^2 r$, $f'(r_h) = -2H$ giving rise to $\kappa = H$.

The associated Hawking temperature is therefore

$$T = \frac{\kappa}{2\pi} = \frac{\hbar H}{2\pi}, \tag{11}$$

where the Boltzmann constant is $k_B = 1$ in our unit convention.

Accordingly, we have a Hawking temperature of the form

$$T = \frac{1}{2\pi}\sqrt{\frac{\Lambda}{3}}. \tag{12}$$

At this stage, the entropy of any Killing horizon equals one quarter of its area,

$$S = \frac{A}{4}, \tag{13}$$

https://doi.org/10.3390/e28010043

leading, for de Sitter, to

$$S = \frac{A}{4} = \frac{\pi}{H^2} = \frac{3\pi}{\Lambda}. \tag{14}$$

Euclidean Derivation for the Temperature

The same result follows from imposing regularity of the Euclidean metric via a Wick rotation, namely $t \to -i\tau$, obtaining $ds_E^2 = f(r)\,d\tau^2 + \frac{dr^2}{f(r)} + r^2 d\Omega_2^2$. Near the horizon, this metric can be approximated as follows:

$$f(r) \approx f'(r_h)(r - r_h) = -2H(r - r_h). \tag{15}$$

An appropriate coordinate change can thus be used to recast the de Sitter metric into a Rindler spacetime. So, defining ρ via

$$d\rho^2 = \frac{dr^2}{f(r)} \approx \frac{dr^2}{-2H(r - r_h)}, \tag{16}$$

and

$$f(r) \approx 2H(r_h - r) \approx H^2\rho^2, \tag{17}$$

we conclude that the metric near the horizon becomes

$$ds_E^2 \approx \rho^2 H^2\, d\tau^2 + d\rho^2 + r_h^2 d\Omega_2^2. \tag{18}$$

Hence, the (ρ, τ) plane is regular only if τ has period

$$\beta = \frac{2\pi}{H}, \tag{19}$$

which yields

$$T = \frac{1}{\beta} = \frac{H}{2\pi}, \tag{20}$$

in agreement with the surface gravity calculation.

2.2. *Specific Heat of de Sitter Space*

The thermodynamic behavior of de Sitter space can be further analyzed by studying the *specific heat* associated with the cosmological horizon. In semiclassical gravity, one can treat the horizon as a thermodynamic system characterized by entropy S, temperature T, and an effective energy E.

The *specific heat* is defined as

$$C \equiv \frac{\partial E}{\partial T} = T\frac{\partial S}{\partial T}, \tag{21}$$

where we express $dE = TdS$ as a function of T by eliminating H. From $T = H/(2\pi)$, we have

$$H = 2\pi T, \qquad r_h = \frac{1}{H} = \frac{1}{2\pi T}, \tag{22}$$

indicating that

$$S = \frac{1}{4\pi T^2} \tag{23}$$

This result is consistent with the fact that $S > 0$.

Substituting into the expression for the specific heat, we obtain

$$C = -\frac{1}{2\pi T^2}. \tag{24}$$

https://doi.org/10.3390/e28010043

The result indicates, this time, that $C \equiv -2S$.

One arrives at the same result, noticing that $C \equiv T\frac{\partial S}{\partial r_h}\left(\frac{\partial T}{\partial r_h}\right)^{-1}$.

The negative sign of C shows that the de Sitter cosmological horizon, like the Schwarzschild black hole horizon, has a *negative specific heat*. This indicates that the thermodynamic system is *thermodynamically unstable in a canonical ensemble*, as an increase in energy reduces the temperature.

In terms of the cosmological constant, Λ, we obtain

$$T = \frac{1}{2\pi r_h}, \qquad C = -\frac{1}{2\pi}\cdot(2\pi r_h)^2 = -2\pi r_h^2 = -\frac{6\pi}{\Lambda}. \tag{25}$$

This result emphasizes the similarity between de Sitter horizons and black hole horizons: both exhibit thermodynamic instability in canonical ensembles, as expected for self-gravitating systems.

One may ask whether this result is general, or, in other words, whether the above occurrence is a direct consequence of the spherical symmetry alone. In addition, one may wonder under which circumstances a spherically symmetric object could exist without a central mass, M, supported only by a cosmological constant. The presence of mass, in fact, is essential for characterizing known compact objects, while de Sitter or generalized de Sitter space contain only Λ, without an associated central mass.

In general, recent observations have not revealed objects that have exact spherical symmetry, instead indicating the presence of quadrupole moments or angular momentum [30]. The *Kerr hypothesis* states that compact objects are described by either Kerr or Schwarzschild metrics. However, thermodynamic instability appears to be a symmetry-related effect and suggests, that at least within general relativity, compact objects endowed with exact spherical symmetry may not truly exist in nature. On the other hand, observations seem to indicate that astrophysical compact objects are rotating, and therefore may point toward excluding those solutions that exhibit negative specific heats. Hence, although spherical symmetry remains essential from a theoretical viewpoint, more studies are needed in light of the possible thermodynamic instability that may preclude the existence of purely spherically symmetric compact configurations.

Below, we therefore deform the de Sitter space, computing the corresponding thermodynamics.

3. Generalization of de Sitter Space

Generalized de Sitter space or, more briefly, quasi-de Sitter space, can be argued following the argument raised in Ref. [31,32].

The corresponding metrics can be obtained in two different ways, as reported below.

- Extending de Sitter space. This procedure requires finding *the most general metric that emulates de Sitter, providing isotropic pressures that are not constant, as well as an energy density that is a function of the radial coordinate*.
- Requiring a finite region with an emergent Cauchy horizon. In this way, it is possible to construct a region analogous to the de Sitter space, in which an observer placed in the center of the reference frame encounters a Cauchy horizon, which does not exist in pure de Sitter space.

For the first approach, we can consider a spherically symmetric spacetime [33]

$$ds^2 = -f(r)\,dt^2 + \frac{1}{g(r)}\,dr^2 + h(r)\,d\Omega^2, \tag{26}$$

https://doi.org/10.3390/e28010043

whose stationary region, where the Killing field ∂_t is timelike, corresponds to the domain in which all three functions remain positive. In this coordinate chart (t, r, θ, φ), the energy–momentum tensor of an anisotropic fluid takes the diagonal form

$$\left[T^{\mu}_{\nu}\right] = \begin{bmatrix} -\rho & 0 & 0 & 0 \\ 0 & P_r & 0 & 0 \\ 0 & 0 & P_t & 0 \\ 0 & 0 & 0 & P_t \end{bmatrix}, \tag{27}$$

where ρ denotes the energy density, while P_r and P_t represent the radial and tangential pressures, respectively.

Imposing spherical symmetry, one may always redefine the radial coordinate such that $h(r) = r^2$, which is exactly the case that we consider below. The resulting general metric describes the spherically symmetric solution of an anisotropic fluid, in which we intend to impose isotropic pressure and include a cosmological constant as part of the energy–momentum tensor, although it is quite different from the de Sitter case.

The sign of Λ plays a crucial role in determining whether the spacetime is asymptotically de Sitter ($\Lambda > 0$) or anti-de Sitter ($\Lambda < 0$), with well-known implications for quantum field theory, holography, and the AdS/CFT correspondence [34].

This correspondence asserts a duality between a $(d+1)$–dimensional gravitational theory defined on an asymptotically anti-de Sitter (AdS) spacetime and a d–dimensional conformal field theory (CFT) living on its boundary.

More precisely, the duality asserts an exact equivalence between the partition functions $Z_{\text{grav}}[\phi_0] = Z_{\text{CFT}}[J]$, where ϕ_0 denotes the boundary value of a bulk field ϕ, acting as a source J for the corresponding operator $\mathcal{O}$ in the CFT. Although exploring this is far beyond the scope of the present work, we leave open the possibility of extending the correspondence to a space that resembles the AdS, such as the one that we report below. In other words, our deformed AdS mass may provide a framework for generalizing the correspondence, further motivating the exploration of ways to extend standard AdS and dS frameworks via additional metrics that look similar to the original formulation.

Black hole thermodynamics in AdS space, for instance, exhibit the Hawking–Page transition.

In our framework, we focus exclusively on positive Λ to ensure the existence of an horizon and, accordingly, a thermodynamic framework, and we require that *corresponding metric asymptotically resembles quintessence.*

Accordingly, for quintessence, additional restrictions over the equation of state are needed. Specifically, defining $w \equiv \frac{P}{\rho}$ and having $P_r = P_t = P$, we require that asymptotically $-1 < w < 0$.

Motivated by this framework, it is natural to consider the following:

- a class of regular, spherically symmetric spacetimes allowing for a vacuum energy contribution capable of inducing a de Sitter phase with a (not necessarily constant) effective energy density Λ;
- an equation-of-state parameter w approaching a constant asymptotic value, consistent with experimental constraints [35,36], thereby enabling the existence of a dark energy component that may violate the WEC.

Incorporating a deformation of the standard de Sitter space, we can write

$$ds^2 = -\left(1 - \frac{\Lambda}{3} r^{2+\epsilon}\right) dt^2 + \frac{1}{g(r)} dr^2 + r^2 d\Omega^2, \tag{28}$$

https://doi.org/10.3390/e28010043

where deviations from the pure de Sitter geometry are encoded in ϵ and in the requirement to have isotropic but nonconstant pressures. Specifically, $g(r)$ is chosen to be regular at $r = 0$ and to grow as $g(r) \sim r^2$ in the limit $r \to \infty$, thus restoring a de Sitter asymptotic regime.

Thus, from the Einstein field equations, we end up with

$$g(r) = f(r)\,\frac{{}_2F_1\left(-\frac{2}{\epsilon+2}, -\frac{\epsilon}{\epsilon+4}; \frac{\epsilon}{\epsilon+2}; \frac{\Lambda}{6}(\epsilon+4)r^{\epsilon+2}\right) + k_0 r^2}{\left(1 - \frac{\Lambda}{6}(\epsilon+4)r^{\epsilon+2}\right)^{\frac{\epsilon}{\epsilon+4}+1}}, \tag{29}$$

where ${}_2F_1$ is the analytical continuation of the hypergeometric function and k_0 a further integration constant, whose meaning will be clarified later in the text. The above expression turns out to be particularly interesting in the case $\epsilon = 0$ and $k_0 \neq -\frac{2}{3}\Lambda$, giving rise to

$$ds^2 = -\left(1 - \frac{\Lambda}{3}r^2\right)dt^2 + \frac{(1 - \frac{2}{3}\Lambda r^2)dr^2}{\left(1 - \frac{\Lambda}{3}r^2\right)(1 + k_0 r^2)} + r^2 d\Omega_2^2, \tag{30}$$

where the significance of the integration constant k_0 is not known *a priori*. Indeed, in the original formulation of the metric, see Ref. [31], no definitive conclusions were drawn concerning this constant. There, it was argued that, when $\epsilon \to 0$, de Sitter solution is not recovered *not because of the presence of* k_0 but rather because k_0 is *a priori* unbounded and may significantly depart from the case $k_0 = -\frac{2}{3}\Lambda$; this is in fact required to recover de Sitter space. Thus, in this work, we aim at finding a possible connection between this further constant, k_0, and the underlying thermodynamics of our extended de Sitter space.

Concerning the second approach, the above metric can be derived alternatively from the previously obtained solution, constructed by starting from a spherically symmetric, singular seed metric subjected to an additive deformation, as proposed in Ref. [37].

We assume that the metric components can be expressed via a δ–expansion, where δ parameterizes deviations from the vacuum Schwarzschild–de Sitter solution. Thus, ensuring $\chi(r) = 1 - \frac{2M}{r} - \frac{\Lambda}{3}r^2$, we find explicitly $f(r) = \chi(r) + \delta f_1(r) + \mathcal{O}(\delta^2)$ and $g(r) = \chi(r) + \delta g_1(r) + \mathcal{O}(\delta^2)$.

Accordingly, the same class of solutions previously analyzed can be recovered by starting from a spherically symmetric singular background and introducing an additive deformation, following the approach of Ref. [37]. To this end, we assume that the metric functions admit a δ–expansion, where the parameter δ quantifies the deviation from the vacuum solution with a cosmological constant, i.e., from the Schwarzschild-de Sitter spacetime. Accordingly, we write

$$f(r) = \chi(r) + \delta\, f_1(r) + \ldots, \tag{31}$$

$$g(r) = \chi(r) + \delta\, g_1(r) + \ldots, \tag{32}$$

with $\chi(r) = 1 - \frac{2M}{r} - \frac{\Lambda}{3}r^2$. The isotropy condition can be considerably simplified by introducing the additional ansatz

$$f_1(r) = \chi(r)\,\alpha(r), \qquad g_1(r) = \chi(r)\,\beta(r). \tag{33}$$

Focusing on the *exact linear solutions*, obtained by setting $\alpha(r) = 0$, one finds

$$f(r) = \chi(r), \qquad g(r) = \chi(r)\big(1 + \delta\,\beta(r)\big), \tag{34}$$

https://doi.org/10.3390/e28010043

so that the isotropy condition reduces to the ordinary differential equation

$$\frac{\beta'(r)}{\beta(r)} = -\frac{6}{3M + 2\Lambda r^3 - 3r}. \tag{35}$$

In this framework, the energy–momentum tensor turns out to be exactly linear in δ, namely

$$T^0_{\ 0} = -\Lambda + \left(-\Lambda + \frac{2}{r^2} - \frac{3(r-3M)}{r^2(3M + 2\Lambda r^3 - 3r)}\right)\delta\,\beta(r), \tag{36}$$

$$T^i_{\ i} = -\Lambda + \left(-\Lambda + \frac{1}{r^2}\right)\delta\,\beta(r), \tag{37}$$

where $\beta(r)$ is the solution of Equation (35), which is manifestly a linear differential equation. At this stage, then, we can take the parameter δ to be absorbed into the integration constant (denoted by D) emerging from Equation (35) by defining $\Upsilon := D\,\delta$. One is then left with a family of solutions parameterized by the single real constant Υ.

In general, the metric functions $f(r)$ and $g(r)$ obtained in this way do not necessarily share the same sign for all $r > 0$, a condition that must be enforced in order to ensure a Lorentzian geometry, as discussed earlier.

Assuming $M > 0$ and $\Lambda < 1/(9M^2)$, both $f(r)$ and $g(r)$ are positive in a neighborhood (r_+, r_c), where r_+ denotes the black hole horizon and r_c the cosmological horizon. Furthermore, for $\Upsilon > 0$, one can show the existence of a radius $r_0 \in (r_+, r_c)$ at which curvature invariants diverge. In this case, the static geometry is well defined only in the interval (r_0, r_c) and exhibits a naked singularity as $r \to r_0$.

For $\Upsilon < 0$, the structure of the geometry is more involved, since $g(r)$ vanishes at a radius $r_{\text{thr}} \in (r_+, r_c)$, although the metric remains regular at that location.

If $M = 0$, one recovers our initial ansatz for $f(r)$,

$$f(r) = 1 - \frac{\Lambda}{3}r^2, \tag{38}$$

which coincides with our solution in the limit $\epsilon = 0$.

This justifies our initial choice of $f(r)$, made with the aim of finding $g(r)$.

It can also be shown that, under the above assumptions on Λ and Υ, the metric is defined for all $r \in (r_{\text{thr}}, r_c)$. By gluing two copies of such a solution at $r = r_{\text{thr}}$, one can construct a wormhole geometry, as discussed in Ref. [37].

In summary, our metric is sufficiently general to reproduce both singular and regular solutions. When additional requirements are imposed, it also admits wormhole configurations.

4. Thermodynamics of Quasi-de Sitter Space

Below, we consider the special case of Equation (30) as a starting point, employing $\epsilon = 0$. The effects of deformation, induced by ϵ, have been discussed in Ref. [31] and appear to not affect the spacetime structure.

Since $f \neq g$, the surface gravity, κ turns out to be

$$\kappa = \frac{1}{2}\sqrt{|f'(r_h)g'(r_h)|}, \tag{39}$$

and, so, the Hawking temperature becomes, in this case,

$$\tilde{T} = \frac{\kappa}{2\pi} = \frac{1}{4\pi}\sqrt{|f'(r_h)g'(r_h)|}, \tag{40}$$

https://doi.org/10.3390/e28010043

where the tilde˜indicates that this temperature is, in principle, different from T, which was previously introduced.

Accordingly, the specific heat becomes

$$\tilde{C} = \tilde{T}\frac{d\tilde{S}}{d\tilde{T}}, \tag{41}$$

under the assumption that k_0 is fully *independent of* Λ. This immediately yields

$$\tilde{S} = S, \tag{42}$$

$$\tilde{T} = \frac{1}{2\pi}\sqrt{\left|k_0 + \frac{1}{r_h^2}\right|}, \tag{43}$$

$$\tilde{C} = -2\pi\left|k_0 + \frac{1}{r_h^2}\right| r_h^4, \tag{44}$$

which, in the limiting case, $k_0 \to -\frac{2}{3}\Lambda$ reduces to $\tilde{T} \to T$, $\tilde{S} \to S$ and $\tilde{C} \to C$.

Alternatively, the specific heat is given by

$$\tilde{C} = -\frac{8\pi^3 T^2}{(k_0 - 4\pi^2 T^2)^2}, \tag{45}$$

where $k_0 < 0$ is an unavoidable condition, thereby forcing $\tilde{C} < 0$.

A similarly intriguing outcome arises when expressing $\tilde{C}$ as

$$\tilde{C} = -6\pi\left(\frac{|3k_0 + \Lambda|}{\Lambda^2}\right), \tag{46}$$

At first glance, there is no way to ensure thermodynamic stability.

Nevertheless, if $k_0 = k_0(r_h)$, then $[k_0] \equiv [\Lambda] \equiv [L]^{-2}$.

Hence, to guarantee that the metric g does not change its sign, i.e., that the Lorentzianity is preserved, we require

$$\frac{1 + k_0 r^2}{1 - \frac{2}{3}\Lambda r^2} > 0, \tag{47}$$

since before the horizon, r_h, $1 - \frac{\Lambda}{3}r^2 > 0$. Moreover, around $r \simeq 0$, expanding to the second order gives

$$g \sim 1 + \left(k_0 + \frac{\Lambda}{3}\right)r^2 + \ldots, \tag{48}$$

Therefore, without loss of generality, for a positive cosmological constant, the solution cannot be defined for all $r > 0$ if our solution is not a de Sitter space.

Hence, for $k_0 \leq -2\Lambda/3$, we expect to have $r \in]0, 1/\sqrt{-k_0}[\cup]\sqrt{3/(2\Lambda)}, +\infty[$, corresponding to two patches. The first, $r \in]0, 1/\sqrt{-k_0}[$, is regular up to $r = 0$, but with no horizon and, accordingly, no associated thermodynamics. The second patch has a singular boundary placed at $r = \sqrt{\frac{3}{2\Lambda}}$. In all other cases, the singularity is present and contained in the stationary region bounded by the horizon $r_h = \sqrt{\frac{3}{\Lambda}}$.

In addition, the case $-2\Lambda/3 < k_0 < -\Lambda/3$ provides a solution, defined within $r \in]0, \sqrt{3/(2\Lambda)}[\cup]1/\sqrt{-k_0}, +\infty[$.

The first patch is again regular, featuring a singular noncentral boundary at $r = \sqrt{3/(2\Lambda)}$. The other patch has an horizon but remains regular up to the boundary $r = 1/\sqrt{-k_0}$.

https://doi.org/10.3390/e28010043

Last but not least, for $-\Lambda/3 \leq k_0 < 0$ appears similar to the above case, with the exception that the outer patch $r \in]1/\sqrt{-k_0}, +\infty[$ has no horizons. Finally, for $k_0 \geq 0$, the outer patch disappears, leaving only the patch defined in the right neighborhood of $r = 0$ given by $r \in]0, \sqrt{3/(2\Lambda)}[$, exists, which again contains a noncentral singularity.

All these considerations, together with the expansion of g around $r \simeq 0$, suggest that the overall thermodynamics in the generalized de Sitter spacetime are not altered significantly, and the presence of k_0 appears as a shift in Λ.

Accordingly, as k_0 is negative, its primary effect is to reduce the specific heat without substantially modifying the overall thermodynamics, as stated above.

Last but not least, one can wonder how using Equation (29), instead of Equation (30) affects the specific heats. Since ϵ is very small, it does not influence the properties of the specific heat and, as has been shown in Ref. [31], its effect is not thermodynamic, so the overall results are largely unchanged and the specific heats remain negative.

4.1. Euclidean Derivation for the Temperature

The same procedure can be applied using the Euclidean formulation. Since our metric in Equation (30) satisfies $f(r) \neq g(r)$, our above treatment must be extended to this case, as we show below.

Near the horizon, we expand both the functions through

$$f(r) \overset{r \to r_h}{\approx} f'_h(r - r_h), \qquad g(r) \overset{r \to r_h}{\approx} g'_h(r - r_h), \tag{49}$$

with $f'(r_h) \equiv f'_h$ and $g'(r_h) \equiv g'_h$, yielding, near the horizon,

$$ds^2 \approx -\frac{f'_h z}{f_0} d\tau^2 + \frac{1}{g'_h z} dz^2, \tag{50}$$

where we have used the fact that, for our metric, $f(r_h) = g(r_h) = 0$. Moreover, we set $d\Omega = 0$ for simplicity and $r = r_0$ as a static point, where the observer measures proper time, so that $d\tau = \sqrt{f_0}\, dt$, with $f_0 = f(r_0)$.

Through a coordinate transformation, $dR \approx \frac{1}{\sqrt{g'_h z}} dz$, we obtain

$$ds^2 \approx -\frac{f'_h g'_h}{4 f_0} R^2 d\tau^2 + dR^2. \tag{51}$$

Following the same procedure used in Equation (19), we find here

$$\tilde{T}^2 (2\pi)^2 = \frac{f'_h g'_h}{4 f_0} \rightleftharpoons \tilde{T} = \frac{\sqrt{|f'_h g'_h|}}{2\sqrt{f_0}(2\pi)}, \tag{52}$$

This reduces to Equation (40) if $f_0 = 1$, certifying the viability of our analysis.

4.2. Comparative Remarks and Cosmological Implications

The quasi-de Sitter construction developed in this work naturally invites comparison with many modified gravity scenarios, including $f(R)$ extensions, nonlinear electrodynamics (NLED) models, and semiclassical quantum-gravity corrections. There, the temperature and entropy of cosmological or black-hole horizons acquire model-dependent contributions. For instance, in NLED-based regular geometries [21], as well as in anti-de Sitter-inspired extensions and deformed black-bounce metrics, the thermodynamic quantities typically depend on additional curvature or on matter couplings that alter the effective gravitational sector.

https://doi.org/10.3390/e28010043

In contrast, our solution relies on general relativity alone, as the metric modification enters exclusively through a geometrical integration constant k_0, without introducing higher-order curvature invariants or new propagating degrees of freedom.

Moreover, the case $k_0 = 0$ is, at least in principle, possible. This provides a clear benchmark by which to disentangle genuine nongeneral relativity thermodynamic effects from those induced simply by modifying the radial dependence of the metric functions.

Our analysis shows that, even when the metric is not written in Schwarzschild coordinates, the resulting thermodynamics remain formally identical to the de Sitter case once k_0 is re-expressed in terms of Λ.

We thus interpret the role of k_0 as an additional cosmological constant, which, however, does not influence the net thermodynamics.

This behavior may significantly differ across alternatives to general relativity [38,39], where additional couplings typically produce nonuniversal corrections to the horizon temperature or entropy. Although our treatment is analytical, the behavior of thermodynamic quantities can be complemented with a numerical exploration of the parameter space (Λ, k_0), as can be found in Ref. [31].

From a cosmological perspective, the invariance of the entropy-specific heat structure indicates that the quasi-de Sitter phase cannot mimic dark energy models featuring evolving effective vacuum energy, such as those driven by K-essence [40] or chiral-quintom fields [41].

Instead, the model represents a geometric deformation of de Sitter space that preserves the intrinsic thermodynamic signature of vacuum-dominated cosmologies.

This reinforces the interpretation that the observed thermodynamic instability is not an artifact of the deformation, but rather a robust feature of spherically symmetric general relativity backgrounds with a cosmological horizon.

4.3. Physical Interpretation of Negative Specific Heat and Thermodynamic Instability

The negative specific heat in our quasi-de Sitter configuration has important physical implications in cosmology, see e.g., [42].

More broadly, in gravitational systems, this property is commonly interpreted as signature for a self-gravitating instability: *an increase in internal energy leads to a decrease in temperature, preventing the construction of a canonical thermodynamic ensemble.*

This feature is quite common since, as stated above, it is even shared by the simplest Schwarzschild solution, as well as by more complicated spacetimes.

Within the present framework, the negative sign of the specific heat is not altered by k_0, which represents the parameter responsible for deviations from a de Sitter space.

This indicates that this instability is structurally tied to the combination of spherical symmetry and the existence of a cosmological horizon, although no theorem, so far, has mathematically demonstrated this.

The generalized de Sitter space modifies the local relation between (f, g) in de Sitter space but leaves the global thermodynamic behavior almost unchanged, suggesting that the instability is controlled primarily by the vacuum-energy term, namely Λ.

In quantum gravity and in approaches inspired by it, such as semiclassical corrections, the negative specific heat is sometimes interpreted as a signal of missing degrees of freedom in the coarse-grained thermodynamic description, which is clearly fully classical, as noted above in the context of the AdS/CFT correspondence.

Similarly, modified gravity models can map this property through the presence of additional curvature modes that redistribute energy nonlocally.

In contrast, our construction does not involve extra dynamical fields and the instability is thus genuinely geometric, originating from the global structure of the horizon itself.

https://doi.org/10.3390/e28010043

These considerations reinforce the conclusion that resolving the thermodynamic instability of de Sitter-like spacetimes may require either

- breaking spherical symmetry in general relativity,
- including rotating or axisymmetric deformations, as prompted in Ref. [32], or
- extending general relativity by introducing additional degrees of freedom.

Moreover, for completeness, the negative behavior of specific heats can be associated with long-range interactions, which may be interpreted as a possible signature of nonlocal effects in gravity. More plausibly, however, the thermodynamic stability may be restored depending on the underlying underlying microphysics, a possibility that remains unclear in the literature.

5. Final Outlooks

In this work, we explored the thermodynamic properties of a quasi-de Sitter spacetime obtained by deforming the standard de Sitter space.

Specifically, we introduced two distinct procedures to derive the metric under consideration and provided two complementary interpretations of the resulting solution. We demonstrated that this geometry exhibits an asymptotically quintessence-like behavior, characterized by isotropic pressures and a nonconstant energy–momentum tensor and that it thereby differs substantially from the exact de Sitter configuration.

Moreover, we highlighted that a nonvanishing cosmological constant—or, equivalently, a vacuum energy contribution—persists in this scenario, with the metric admitting a finite radial domain wherein the Lorentzian signature is preserved.

A key feature of this construction is the emergence of an additional parameter, denoted k_0, whose thermodynamic implications we examined in detail. Even though k_0 is naively found by Einstein's field equations, requiring isotropic pressure, its role appears *a priori* unspecified. Moreover, the modifications over the de Sitter phase do not depend on it alone, but also depend on a deformation on the shift function that also involves Λ.

Afterwards, we explicitly derived the entropy, Hawking temperature, and specific heat associated with this geometry, performing a comparative analysis with their de Sitter counterparts.

While the entropy remains unchanged, since the horizons are the same in both spacetimes under consideration, both the Hawking temperature and the specific heat acquire nontrivial corrections, parameterized by k_0. At first glance, these deviations appear to increase the complexity of the deformed metric; however, by interpreting k_0 as a parameter of the same dimensionality as the cosmological constant and imposing $k_0 < 0$ and $k_0 < \frac{\Lambda}{3}$, we demonstrated that the thermodynamic quantities are slightly modified from those of de Sitter spacetime.

This reinforces the conclusion that the thermodynamic structure is intrinsically determined by the underlying geometric properties encoded in Λ alone, which remains formally identical between the quasi–de Sitter and de Sitter cases.

The issue of negative specific heat, along with its associated inconsistencies and indications of thermodynamic instability, remains here unresolved. Clearly, this depends on how $f(r), g(r)$ and $h(r)$ have been defined throughout our manuscript. Even though none of these quantities has been arbitrarily fixed, as stressed in the text, our metric deforms de Sitter space in a very precise and possibly nonunique manner. Alternatives may be explored in future efforts, although our findings appear consistent with the interpretation that these pathologies are inherently linked to spherical symmetry and to pure general relativity.

Several future directions naturally emerge, as summarized below.

https://doi.org/10.3390/e28010043

(i) Turbulent and geometric instabilities. As discussed in Ref. [29], anti-de Sitter spacetime may exhibit weakly turbulent instabilities, where nonlinear mode coupling drives the system toward collapse. Given the similarity between our quasi-de Sitter model and de Sitter space, a full nonlinear stability analysis, combining perturbative techniques and numerical evolution, would be a useful future step.
(ii) Axisymmetric and rotating generalizations. Since the thermodynamic instability persists under spherical symmetry, a natural extension involves relaxing this symmetry and constructing axisymmetric solutions [32]. There, the role of the parameter k_0 would be re-explored, with the expectation that it would significantly differ from the present case.
(iii) Inclusion of matter sources and mass terms. Another promising direction concerns incorporating additional sources into the metric, such as scalar fields, anisotropic fluids or NLED. This prerogative will be investigated to understand how these sources influence the horizon thermodynamics.
(iv) Quantum and semiclassical corrections. The present work is fully classical; therefore, we intend to study semiclassical effects such as trace anomalies, one-loop corrections to the effective action, or nonperturbative quantum-gravity contributions to determine whether these effects can alter the sign of the specific heat.
(v) Euclidean quantum gravity and path-integral extensions. The Euclidean derivation presented here may be generalized by constructing the full Euclidean instanton associated with the quasi-de Sitter metric.
(vi) Implications for cosmology and vacuum-energy modeling. Since the asymptotic behavior of our solution resembles that of quintessence-like configurations, a systematic study of its cosmological embedding would be highly relevant, and, in particular, it would be worthwhile to explore whether the quasi-de Sitter phase may serve as an effective description of transitions between vacuum-dominated epochs, among other studies.

Author Contributions: Writing—original draft, O.L., M.A. and K.B. All authors have read and agreed to the published version of the manuscript.

Funding: O.L. acknowledges support by the Fondazione ICSC, Spoke 3 Astrophysics and Cosmos Observations. National Recovery and Resilience Plan (Piano Nazionale di Ripresa e Resilienza, PNRR) Project ID *CN*00000013 "Italian Research Center on High-Performance Computing, Big Data and Quantum Computing" funded by MUR Missione 4 Componente 2 Investimento 1.4: Potenziamento strutture di ricerca e creazione di "campioni nazionali di R&S (M4C2-19)"—Next Generation EU (NGEU).

Data Availability Statement: The original contributions presented in this study are included in the article. Further inquiries can be directed to the corresponding author.

Conflicts of Interest: The authors declare no conflicts of interest.

References

1. Abbott, B.P.; Abbott, R.; Abbott, T.D.; Abernathy, M.R.; Acernese, F.; Ackley, K.; Adams, C.; Adams, T.; Addesso, P.; Adhikari, R.X.; et al. Observation of Gravitational Waves from a Binary Black Hole Merger. *Phys. Rev. Lett.* **2016**, *116*, 061102. [CrossRef]
2. Arun, K.G.; Belgacem, E.; Benkel, R.; Bernard, L.; Berti, E.; Bertone, G.; Besancon, M.; Blas, D.; Böhmer, C.G.; Brito, R.; et al. New horizons for fundamental physics with LISA. *Living Rev. Relativ.* **2022**, *25*, 4. [CrossRef]
3. Dowker, F.; Gauntlett, J.P.; Kastor, D.A.; Traschen, J.H. Pair creation of dilaton black holes. *Phys. Rev. D* **1994**, *49*, 2909–2917. [CrossRef]
4. Boshkayev, K.; Suliyeva, G.; Ivashchuk, V.; Urazalina, A. Circular geodesics in the field of double-charged dilatonic black holes. *Eur. Phys. J. C* **2024**, *84*, 19. [CrossRef]

https://doi.org/10.3390/e28010043

5. Boshkayev, K.; Takey, G.; Ivashchuk, V.D.; Malybayev, A.N.; Nurbakova, G.S.; Urazalina, A. Stability analysis of circular geodesics in dyonic dilatonic black hole spacetimes. *Phys. Dark Universe* **2025**, *48*, 101862. [CrossRef]
6. Malybayev, A.N.; Boshkayev, K.A.; Ivashchuk, V.D. Quasinormal modes in the field of a dyon-like dilatonic black hole. *Eur. Phys. J. C* **2021**, *81*, 475. [CrossRef]
7. Hughes, S.A. Parameterizing black hole orbits for adiabatic inspiral. *Phys. Rev. D* **2024**, *109*, 064077. [CrossRef]
8. Barack, L.; Long, O. Self-force correction to the deflection angle in black-hole scattering: A scalar charge toy model. *Phys. Rev. D* **2022**, *106*, 104031. [CrossRef]
9. Ashoorioon, A.; Casadio, R.; Jafarzade, K.; Jahani Poshteh, M.B.; Luongo, O. Gravitational radiation reaction around a static black hole surrounded by a Dehnen type dark matter halo. *arXiv* **2025**, arXiv:2509.08569. [CrossRef]
10. Cardoso, V.; Destounis, K.; Duque, F.; Macedo, R.P.; Maselli, A. Black holes in galaxies: Environmental impact on gravitational-wave generation and propagation. *Phys. Rev. D* **2022**, *105*, L061501. [CrossRef]
11. Figueiredo, E.; Maselli, A.; Cardoso, V. Black holes surrounded by generic dark matter profiles: Appearance and gravitational-wave emission. *Phys. Rev. D* **2023**, *107*, 104033. [CrossRef]
12. Luongo, O.; Muccino, M. Speeding up the universe using dust with pressure. *Phys. Rev. D* **2018**, *98*, 103520. [CrossRef]
13. Capozziello, S.; D'Agostino, R.; Lapponi, A.; Luongo, O. Black hole thermodynamics from logotropic fluids. *Eur. Phys. J. C* **2023**, *83*, 175. [CrossRef]
14. Al-Badawi, A.; Shaymatov, S.; Sekhmani, Y. Schwarzschild black hole in galaxies surrounded by a dark matter halo. *JCAP* **2025**, *02*, 014. [CrossRef]
15. Gohain, M.M.; Phukon, P.; Bhuyan, K. Thermodynamics and null geodesics of a Schwarzschild black hole surrounded by a Dehnen type dark matter halo. *Phys. Dark Universe* **2024**, *46*, 101683. [CrossRef]
16. Al-Badawi, A.; Shaymatov, S. Quasinormal modes and shadow of Schwarzschild black holes embedded in a Dehnen-type dark matter halo exhibiting a cloud of strings. *Commun. Theor. Phys.* **2025**, *77*, 035402. [CrossRef]
17. Malafarina, D. Semi-classical dust collapse and regular black holes. *arXiv* **2022**, arXiv:2209.11406. [CrossRef]
18. Boshkayev, K.; Idrissov, A.; Luongo, O.; Muccino, M. Quasiperiodic oscillations for spherically symmetric regular black holes. *Phys. Rev. D* **2023**, *108*, 044063. [CrossRef]
19. Kurmanov, Y.; Boshkayev, K.; Konysbayev, T.; Luongo, O.; Saiyp, N.; Urazalina, A.; Ikhsan, G.; Suliyeva, G. Accretion disks properties around regular black hole solutions obtained from non-linear electrodynamics. *Phys. Dark Universe* **2024**, *46*, 101566. [CrossRef]
20. Boshkayev, K.; Konysbayev, T.; Kurmanov, Y.; Luongo, O.; Muccino, M.; Taukenova, A.; Urazalina, A. Luminosity of accretion disks around rotating regular black holes. *Eur. Phys. J. C* **2024**, *84*, 230. [CrossRef]
21. Bronnikov, K.A. Regular black holes sourced by nonlinear electrodynamics. *arXiv* **2022**, arXiv:2211.00743. [CrossRef]
22. Bronnikov, K.A. Regular black holes as an alternative to black bounce. *Phys. Rev. D* **2024**, *110*, 024021. [CrossRef]
23. Gibbons, G.W.; Hawking, S.W. Cosmological event horizons, thermodynamics, and particle creation. *Phys. Rev. D* **1977**, *15*, 2738. [CrossRef]
24. Dymnikova, I. Vacuum nonsingular black hole. *Gen. Relativ. Gravit.* **1992**, *24*, 235. [CrossRef]
25. Ayón-Beato, E.; García, A. The Bardeen model as a nonlinear magnetic monopole. *Phys. Lett. B* **2000**, *493*, 149. http://arxiv.org/abs/gr-qc/0009077. [CrossRef]
26. Bardeen, J.M. Non-singular general relativistic gravitational collapse. In Proceedings of the Fifth International Conference on Gravitation and the Theory of Relativity (GR5), Tbilisi, Georgia, 9–13 September 1968.
27. Hayward, S.A. Formation and evaporation of nonsingular black holes. *Phys. Rev. Lett.* **2006**, *96*, 031103. [PubMed]
28. Bizoń, P. Is AdS stable? *Gen. Rel. Grav.* **2014**, *46*, 1724. [CrossRef]
29. Bizon, P.; Rostworowski, A. On weakly turbulent instability of anti-de Sitter space. *Phys. Rev. Lett.* **2011**, *107*, 031102. [CrossRef]
30. Cardoso, V.; Pani, P. Testing the nature of dark compact objects: A status report. *Living Rev. Relativ.* **2019**, *22*, 4. [CrossRef]
31. Giambò, R.; Luongo, O. De Sitter-like configurations with asymptotic quintessence environment. *Class. Quantum Gravity* **2024**, *41*, 125005. [CrossRef]
32. Corona, D.; Giambò, R.; Luongo, O. Motion of test particles in quasi anti-de Sitter regular black holes. *Int. J. Geom. Methods Mod. Phys.* **2024**, *21*, 2440019. [CrossRef]
33. Cosenza, M.; Herrera, L.; Esculpi, M.; Witten, L. Some models of anisotropic spheres in general relativity. *J. Math. Phys.* **1981**, *22*, 118–125. [CrossRef]
34. Hubeny, V.E. The AdS/CFT Correspondence. *Class. Quantum Gravity* **2015**, *32*, 124010. [CrossRef]
35. Aviles, A.; Gruber, C.; Luongo, O.; Quevedo, H. Cosmography and constraints on the equation of state of the Universe in various parametrizations. *Phys. Rev. D* **2012**, *86*, 123516. [CrossRef]
36. Luongo, O.; Muccino, M. Model-independent cosmographic constraints from DESI 2024. *Astron. Astrophys.* **2024**, *690*, A40. [CrossRef]

https://doi.org/10.3390/e28010043

37. Molina, C.; Martin-Moruno, P.; Gonzalez-Diaz, P.F. Isotropic extensions of the vacuum solutions in general relativity. *Phys. Rev. D* **2011**, *84*, 104013. [CrossRef]
38. Johnson, C.V. Specific Heats and Schottky Peaks for Black Holes in Extended Thermodynamics. *Class. Quantum Gravity* **2020**, *37*, 054003. [CrossRef]
39. Thomas, B.B.; Saleh, M.; Kofane, T.C. Thermodynamics and phase transition of the Reissner-Nordstroem black hole surrounded by quintessence. *Gen. Relativ. Gravit.* **2012**, *44*, 2181–2189. [CrossRef]
40. Bilic, N. Thermodynamics of k-essence. *Phys. Rev. D* **2008**, *78*, 105012. [CrossRef]
41. Paliathanasis, A. Unified dark energy from Chiral-Quintom model with a mixed potential in Friedmann–Lemaître–Robertson–Walker cosmology. *Eur. Phys. J. C* **2023**, *83*, 756. [CrossRef]
42. Luongo, O.; Quevedo, H. Cosmographic study of the universe's specific heat: A landscape for Cosmology? *Gen. Relativ. Gravit.* **2014**, *46*, 1649. [CrossRef]

https://doi.org/10.3390/e28010043

MDPI AG
Grosspeteranlage 5
4052 Basel
Switzerland
Tel.: +41 61 683 77 34

Entropy Editorial Office
E-mail: entropy@mdpi.com
www.mdpi.com/journal/entropy

www.ingramcontent.com/pod-product-compliance
Lightning Source LLC
Chambersburg PA
CBHW062100220926
51850CB00010B/2839
9783725874408